A Stain Upon the Sea

A Stain Upon the Sea

West Coast Salmon Farming

Stephen Hume, Alexandra Morton,

Betty C. Keller & Rosella M. Leslie,

Otto Langer and Don Staniford

HARBOUR PUBLISHING

Published by
Harbour Publishing Co. Ltd.
P.O. Box 219
Madeira Park, BC
V0N 2H0
www.harbourpublishing.com

Cover design by Peter Read
Page design by Warren Clark

Printed and bound in Canada

Harbour Publishing acknowledges financial support from the Government of Canada through the Book Publishing Industry Development Program and the Canada Council for the Arts, and from the Province of British Columbia through the British Columbia Arts Council and the Book Publisher's Tax Credit through the Ministry of Provincial Revenue.

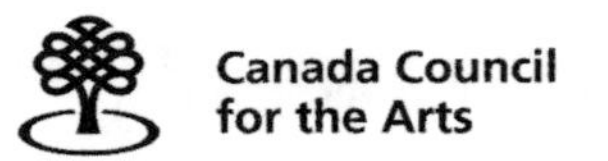

Library and Archives Canada Cataloguing in Publication

A stain upon the sea : west coast salmon farming / Alexandra Morton ... [et al.].

Includes bibliographical references and index.
ISBN 1-55017-317-0

1. Fish-culture—British Columbia. 2. Salmon fisheries—British Columbia. 3. Fish-culture. I. Morton, Alexandra, 1957- II. Title.

HD9469.S23B75 2004a 639.3'756'09711 C2004-904891-0

Acknowledgments

Don Staniford gives special thanks to Allan Berry, Jackie Mackenzie, Jonathan Davis, David Oakes and Deep Trout.

Otto Langer thanks the many environmental groups, journalists, fishermen, academics, citizens and a politician, John Cummins, who have often questioned why the BC and Canadian governments would allow the intrusion of farming salmon into the only place in the world where we have an abundance of wild salmon. Without their public watchdog role and in the face of government abdication of responsibility, salmon farming would be totally out of control on this coast.

Thanks also to Dr. John Volpe, Martin Bühler, Adrian Raeside, Craig Orr and Trish Hall at Watershed Watch Salmon Society, Margo Metcalfe at the David Suzuki Foundation, Howard Silverman at Ecotrust, Jeff Ardron and the Living Oceans Society, Suzanne Connell at the Georgia Strait Alliance, Greg Higgs and Ed May at the Forest Action Network.

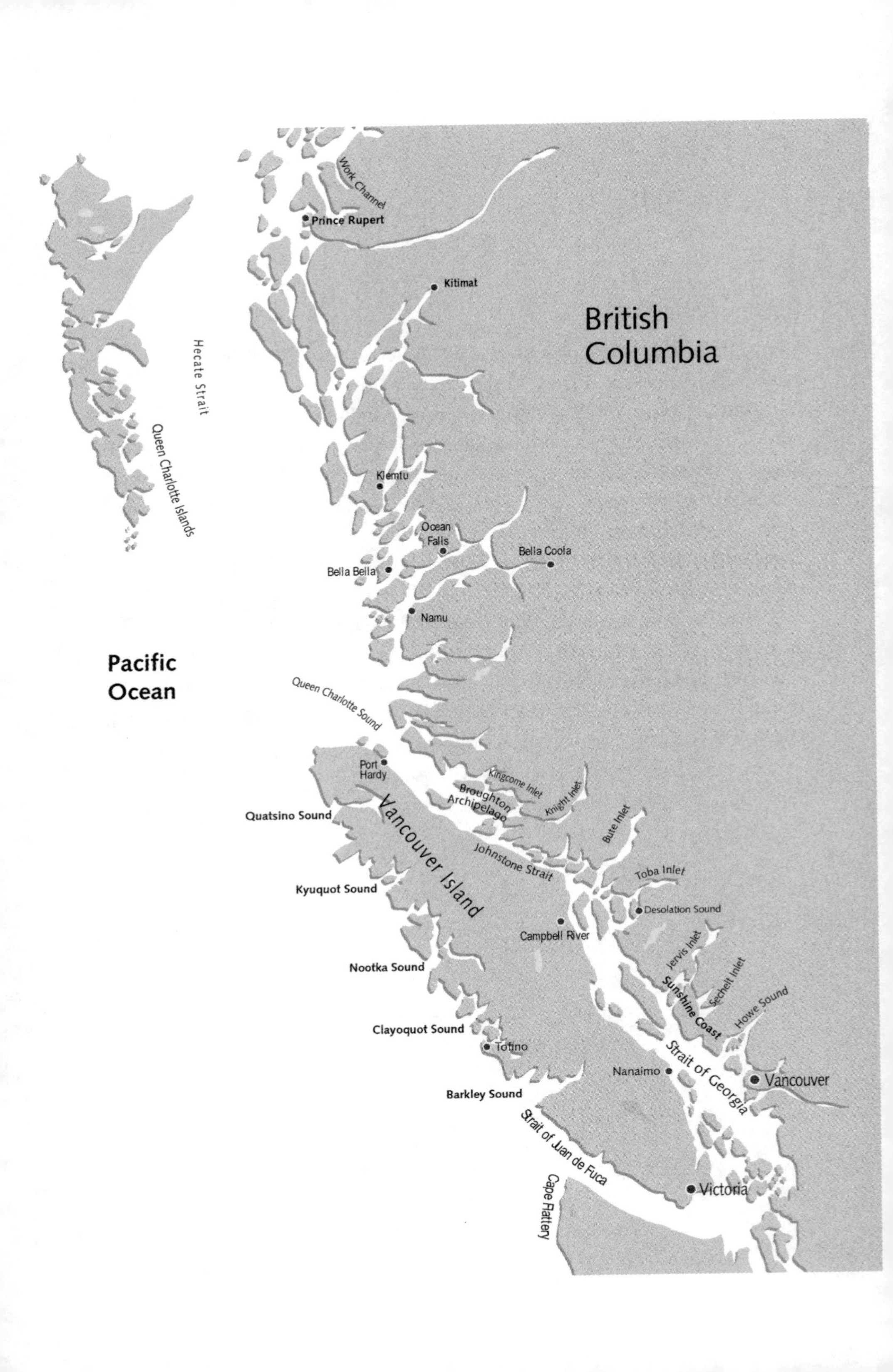

Work Channel
Prince Rupert
Kitimat
British Columbia
Hecate Strait
Queen Charlotte Islands
Klemtu
Ocean Falls
Bella Coola
Bella Bella
Namu
Pacific Ocean
Queen Charlotte Sound
Port Hardy
Kingcome Inlet
Broughton Archipelago
Knight Inlet
Quatsino Sound
Vancouver Island
Bute Inlet
Johnstone Strait
Toba Inlet
Kyuquot Sound
Desolation Sound
Campbell River
Jervis Inlet
Nootka Sound
Sechelt Inlet
Sunshine Coast
Howe Sound
Clayoquot Sound
Tofino
Strait of Georgia
Nanaimo
Vancouver
Barkley Sound
Strait of Juan de Fuca
Cape Flattery
Victoria

Contents

List of Abbreviations

AHDs acoustic harassment devices
AIP Aquaculture Incentive Program
BCP BC Packers Ltd.
BCSFA BC Salmon Farmers Association
BKD bacterial kidney disease
CAAR Coastal Alliance for Aquaculture Reform
CEAA Canadian Environmental Assessment Act
CRIS Coastal Resources Inventory Study
DAFS Department of Agriculture and Fisheries Scotland
DDVP dichlorvos
DFO Department of Fisheries and Oceans
DOE Department of Environment
DOJ Department of Justice
EC European Commission
FoE Friends of the Earth
HABs harmful algal blooms
HEB habitat and enhancement branch (of DFO)
ICES International Council for the Exploration of the Sea
IHN infectious hematopoietic necrosis
MAF Ministry of Agriculture and Food / Ministry of Agriculture and Fisheries
MAFF Ministry of Agriculture, Fisheries and Food / Ministry of Agriculture, Food and Fisheries
MELP Ministry of Environment, Lands and Parks
MLPH Ministry of Lands, Parks and Housing
MoE Ministry of Environment
NDP New Democratic Party
NGO non-governmental organization
OP organophosphate
PAF Pacific Aqua Foods Ltd.
PBS Pacific Biological Station
PFRCC Pacific Fisheries Resource Conservation Council
SAR Salmon Aquaculture Review
SCRD Sunshine Coast Regional District
SEPA Scottish Environmental Protection Agency
TBT tributyltin
TSE Toronto Stock Exchange
UBC University of British Columbia
UFAWU United Fishermen and Allied Workers Union
VHS viral hemorrhagic septicemia
VSE Vancouver Stock Exchange
WHO World Health Organization
WMB Waste Management Branch

Preface

David T. Suzuki

In his seminal book *The Future Eaters,* paleontologist Timothy Flannery points out that the history of our species has been one of movement to new areas where, as an alien arrival, we repeatedly made the same mistakes—over-exploiting abundant and easily collected "resources" such as slow-moving animals like ground sloths, flightless birds and even mammoths. Armed with simple tools and implements, our distant ancestors were clever and effective and succeeded in driving many species to extinction until the land could no longer easily support them. So those paleolithic people disappeared, moved on to repeat the pattern, or learned enough to stay put and live within the productive capacity of their surroundings. According to Flannery, this has been the prehistoric track record of our species.

After the readily habitable areas of Earth had been occupied by humans, a new wave of migration over the past half millennium brought with it a new mindset: to conquer and tame the wilderness, including the indigenous people, and make it over into a familiar and more productive landscape. In only a few centuries this mindset has propelled our species into a dominant position; unfettered by any biological need for specific habitats, we are ubiquitous and now the most numerous mammal on Earth. However, unlike any other species, we have acquired a powerful itch for material goods that is fuelled by a global economy based on generating consumer products and satisfied by unprecedented technological muscle power to extract and process physical and biological material from our surroundings. The consequences have been devastating, creating enormous problems of climate change, toxic pollution, habitat destruction and species extinction, loss of topsoil, and so on.

In the 3.8 billion years that life has existed on Earth, no other species has acquired the power to transform the physical, chemical and biological features of the planet as we do today. As if to underscore this position of dominance, in 1987 the Brundtland Commission released its report, *Our Common Future,* in which it coined the phrase "sustainable development" and suggested that every country should protect twelve percent of its land in its natural state. The tacit assumption of such a target is that we can claim the other 88 percent. Human beings are one species out of an estimated ten to 30 million species, and it is a mite presumptuous to think that all the other species can simply squeeze into the space we leave for them while we use the rest as we please. Aside from the sheer hubris of such a position, it is suicidal because as biological beings, humans remain every bit as dependent as any other species on nature's services for clean air, water, soil and energy.

The planet has been fully occupied and developed by other species for tens of millennia, and the notion that we know enough to be able to push plants and animals around and "manage" natural resources is simply absurd. For one thing, we've learned that while we are indeed clever and can create new and powerful technologies, our knowledge about how everything in the world is intertwined and mutually dependent is minuscule. Over and over we exploit new technologies for immediate benefit, only to learn through catastrophe that there are unexpected repercussions. This cycle has been repeated with fish hatcheries, dams, rapid exploitation of fossil fuels, DDT and other pesticides, nuclear plants, numerous drugs, CFCs, etc. But once an industry is created, jobs, profit and the economy counterbalance the ecological or health hazards and impede quick action. Over and over the vested interests of an industry have retarded efforts to install catalytic converters and airbags in cars, acknowledge tobacco's hazards, reduce pollution by chemical industries, eliminate deadly pesticides, remove lead from gasoline … And when the economy turns down, we act as if nature can increase its productivity when we relax emissions standards, increase allowable catches or bycatch, escalate logging or shipping of raw logs, and so on.

As land-dwelling air breathers, human beings focus on things on solid ground and in the air. When it comes to anything aqueous, our knowledge base is far more limited and fragmentary. It's alien territory to us, and we catch mere glimpses into it with nets, underwater cameras, subs and scuba gear. It's not a surprise, then, that our models of ocean ecosystems are fragmentary and grotesque simplifications. Our terrestrial bias leads us to think that a marine environment is homogeneous and that undersea creatures are replenished as if on an endless conveyor belt. In a civics class in high school back in the

1950s, I remember being taught that the oceans are a source of near-limitless, self-renewing protein. I doubt that was true even fifty years ago, but it was (and still is) the dominant attitude.

Anyone who has fished along the coast of British Columbia over the past half century, or even the last couple of decades, has witnessed the decrease in size and depletion of numbers of sport fish like salmon, snapper and ling, and shellfish like abalone, geoducks and crab. Their rapid decline or disappearance should be an alarm bell that warns us that, as our ancestors did on land, we are exploiting marine resources beyond their replacement capacity and altering their habitat with destructive practices like bottom dragging. If we have anything to learn from the past, it is that nature can be extremely forgiving if we learn to rein in our demands and actions to take the pressure off other species, protect ecosystems and exploit within the reproductive limits of the target species. Wild organisms have programmed behaviours and needs built right into their DNA; they can't simply adapt to human intrusions, like extensive fish farms with open netpens.

Genuine sustainability demands that we recognize and acknowledge some simple principles.

1. **Humility.** Nature has had 3.8 billion years to evolve exquisite strategies for survival. Through careful observation and research we tease out insights into some of those mechanisms. But compared to what we have yet to learn and understand, we have bare-bones knowledge, which means that in a time when we have very powerful but extremely crude technologies, we should maintain the greatest caution in applying them. It would be far better to explore the concept of biomimicry rather than attempting to bludgeon nature into submission.
2. **Long time frames.** We have tended to measure time in terms of a generation or a life, but now the frame has been shortened by the dictates of political pressures set by the interval between elections, the economic bottom lines of annual or quarterly reports, or the media deadlines of breaking news stories. We must think in much longer time frames, dictated by natural cycles rather than the stock market.
3. **Holistic.** We tend to perceive and interact with the world in a fragmented way. Thus, salmon are "managed" by international commissions, federal departments of Indian affairs (Native food fishery) and fisheries and oceans (commercial fishery), and a provincial

department of tourism (sport fishery). As well, trees are controlled by the department of forestry; rivers by energy, agriculture and urban development; mountains and rocks by mining; and eagles, bears and whales by environment. Yet as ecologists like Tom Reimchen have demonstrated, they are all part of a single entity—the forest needs the marine nitrogen concentrated in the bodies of salmon and spread by eagles, bears and insects, while the fish need the trees to cling to the soil, shade the water and provide feed for the fry. We can't manage the components when we deal with the shards of a shattered world piece by piece.

4. **Conservative.** In view of our enormous ignorance of the basis of nature's productivity, we should be very conservative and protect as much of that intact natural state as possible, both for its continued productivity and as a source of knowledge. Being conservative means always erring on the side of caution rather than allowing immediate economic, political or social priorities to set the agenda.

This is the context within which the future of salmon aquaculture should be assessed. Each of the essays in this book is a brilliant exposition of the history, track record and ecological, economic and social impact of this infant activity.

Introduction

Terry Glavin

There is something very old happening here, there is something very new happening here, and there is also something very important happening here. Each of these things must be made plain, straightaway, in any honest account of the story of salmon farming on British Columbia's coast. This book is an attempt to provide just such an account.

Here is what is old about it.

Around the Sea of Okhotsk, and on the mainland side of the Sea of Japan, tribal communities associated with the Yankovskaya culture were engaged in a rudimentary form of oyster breeding about 4,000 years ago. In the San Juan Islands, south of the Strait of Georgia, Strait Salish peoples were engaged in shellfish harvesting at a scale that can easily be described as aquaculture. They were prosecuting a strictly regulated harvest of clams on communally owned tidal flats, protected from winter storms by stone hedges constructed around the low-water mark of bays.

To allow for a certain sad symmetry, one might begin instead in the Broughton Archipelago. In the mists of former times, Kwakwaka'wakw people built more than 300 "clam gardens" buttressed by fourteen kilometres of rock walls, ranging from ten metres to 1,500 metres in length, all in aid of creating textbook-perfect conditions for the optimal growth and density of butter clams, littlenecks, cockles and horse clams. The sad symmetry arises from the fact that in 2001, a sea lice outbreak in the Broughton's salmon farms was followed by one of the most dramatic and rapid collapses of a wild salmon population known to science, when local pink salmon runs fell, in a single generation, from 3.6 million spawners to 147,000 spawners.

Whichever way one chooses to tell the story, it must be said that the origins of aquaculture lie in antiquity. Ancient Egyptians reared tilapia in ponds, and the practice was picked up by the Etruscans and the Greeks. The necessary contemporary preoccupation with aquaculture's prospects for feeding the world might be said to have begun with Aristotle, who considered such things when pondering the spawning habits of minnows. In his *Historia Animalium,* Aristotle wrote: "These fishes shed their eggs little by little . . . If all the eggs were preserved, each species would be infinite in number." Carp ponds were a common feature of the European countryside during the Middle Ages. Almost 3,000 years ago the Chinese were practising intensive carp farming, a form of aquaculture that spread to Indonesia and Malaysia. More than 1,000 years ago the first Hawaiians were raising saltwater fish from enclosed lagoons.

Here's what's new about it.

In the wake of the industrial revolution, the world's human population exploded, and wild fish populations began to collapse around the world. In the final two decades of the 20th century, the world's aquaculture production tripled, and farmed fish came to account for more than a quarter of all fish consumed by human beings. By the dawn of the 21st century, more than 200 species were being farmed in one way or another, from giant clams to the smallest of shrimp.

But that is about pace and scale. It is not what is truly new about what's going on. After all, more than 80 per cent of the world's aquacultural output, in the 21st century's first decade, was still comprised of the species that were there at the beginning. It was still mainly about oysters, clams, tilapia, catfish, and carp.

But the fastest growing sector of the aquaculture industry, worldwide, is salmon farming. And while one might say that pace and scale are not what's new about salmon farming, it is worth reflecting on a couple of things. By the 1990s, farmed salmon outstripped global production of wild salmon. By 2004, global farmed salmon production had reached one million metric tonnes, annually, with British Columbia ranking fourth in production behind Norway, Chile and the United Kingdom. The scale is an important thing to reflect upon. The amount of farmed salmon on the market every year is now greater than the combined human population of British Columbia, Alberta, Washington and Oregon.

But what is truly new about salmon farming is that it represents a departure in the relationship between people and animals arguably as significant as the epochal change ushered in by the neolithic revolution more than 10,000 years ago. Salmon are carnivores. For the first time in history, we are raising carnivores for food. True, the ancient Incas raised dogs in pens for meat, to satisfy the delicate palates of their kings. Also, Russian fur farms have long relied on whale meat to feed their mink. But never before have humans raised carnivores on a mass scale, for mass consumption, as food.

So it is an experiment, one might say. And by so intensively interfering in the process of natural selection, by subjecting salmon to such elaborate methods of artificial selection, by genetic tinkering and by long-term selective breeding, we are creating a wholly new species. Most farmed salmon owe their origins to *Salmo salar,* the wild Atlantic salmon. Most farmed salmon, argues University of Toronto zoologist Mart Gross, should be understood as a new species entirely. Gross calls it *Salmo domesticus.* It is properly understood as an obligate parasite species, like cows or chickens. Not quite wild, not quite tame, it is a species that began as a function of human endeavour. But as with all forms of life, its trajectory cannot be so easily decided.

By 1995, between a quarter and a half of all the salmon feeding in the krill-rich waters of the Northeastern Atlantic were *Salmo domesticus,* escapees from fish farms in Europe and in Canada. It had invaded the spawning habitat of *Salmo salar* from the River Vosso in Norway to the Magaguadavic River in New Brunswick. It had spread far beyond its ancestors' range, into seas that had long been the exclusive haunts of *Salmo salar*'s distant cousins, the six *Oncorhynchus* species of Pacific salmon. It has been found attempting to spawn in at least 50 British Columbia rivers. It has turned up in trawl nets in the Bering Sea.

Even if we succeed in confining this animal to its pens, it is a voracious creature, and salmon farming requires at least three kilograms of wild fish from the high seas to produce every kilogram of farmed salmon. Four of the top five high-seas fisheries in the world now contribute fish food for the aquaculture industry in the form of herring, mackerel, anchoveta, anchovy and sardinella.

Still, it is too easy to demonize salmon farmers. It should not be forgotten that the industry has been encouraged, subsidized and privileged by public policy under governments of every mainstream Canadian political party.

It must also be said that the world is hungry. There are six billion of us, and more than a billion of us rely on fish as our primary source of animal protein. A third of all the marine protein humans consume now comes from aquaculture. The world's traditional fisheries are collapsing, and it isn't just because of unenlightened Third World poacher states or unbridled corporate greed. With the collapse of the North Atlantic cod fishery in the early 1990s, Canada, the birthplace of Greenpeace, was forced to admit that it had utterly destroyed the oldest and largest pelagic fishery in the history of the human experience on earth.

And the world is getting hungrier. There may be as many as twelve billion mouths to feed by the end of the 21st century. The world's oceans have given up all they can give of the high trophic-level fish that human beings prefer. The world needs fish. That fish is going to have to come from somewhere.

So it is an old story, and a new story, and it's also very complicated.

But it is not just about food. The people who eat farmed salmon are not the people who are haunted by the spectre of hunger. The market for farmed salmon is largely among the affluent.

It is not just about money, either. It is doubtful whether the net economic gains associated with direct and indirect salmon farm employment in British Columbia have offset the economic losses associated with the market price distortions caused by farmed salmon, which have driven wild salmon prices paid to fishermen to 1930s-era levels.

It is about something even more important.

It is about salmon.

In Quebec it is about language. In British Columbia it is about salmon. In such matters, people from away must be forgiven for failing to understand why this story is so important. It is about a creature of vital economic and totemic significance to aboriginal cultures that constructed from its flesh a quality of life, and a density of human population, higher than any aboriginal civilization west of the Mississippi and north of the Valley of Mexico.

It is about a creature that has come to swim at the vortex of settler cultures from northern California to southeast Alaska. It is about a "keystone species" that makes of its own bodies the most significant contribution to terrestrial ecological functioning, from woodmites to eagles, from voles to bears, and to the fecundity of the forest itself. Salmon have become a "keystone species" for this part of the world in ways that the ecologists who first coined the term never even imagined.

Anything that interferes with, threatens, infects or disrupts wild Pacific salmon necessarily gives rise to questions about what we ultimately expect of one another as citizens, as British Columbians, as Canadians. It is a complicated debate. This book is an honest contribution to that debate. It is long overdue.

Fishing for Answers

Stephen Hume

For more than 10,000 years, since the last Ice Age clad what is now British Columbia with glaciers two kilometres thick, silvery hordes of salmon have been returning to BC's rivers to sustain a cycle of life and death, abundance and depletion, that has come to characterize not only their vital ecological niche but also the cultural soul of the province itself.

In the fall their vast numbers sustain the great carnivores and predators of western North America. Grizzly bears, the world's largest black bears, wolves, eagles, ospreys, ravens and a host of gulls feast on salmon in seasonal preparation for the lean months of winter. Their rotting carcasses fertilize the riparian fringe, flushing nutrients into the rivers and streams to support aquatic life that in turn supports other species. In spring the millions of alevins, fry and smolts that migrate to the sea provide abundant feed for creatures higher on the food chain, from cutthroat trout to seabirds. In the ocean they are preyed upon by whales, seals and sea lions—and human beings.

The aboriginal peoples the salmon sustained called them "Runners to the Sea" and celebrated their annual return with reverence and special rites. To science they are in many ways still a mystery. To commercial interests they are "pieces," the money fish, mortgage payments and cash in the bank. To sport anglers they remain kings of the light-tackle game fish. No one in British Columbia disputes the assertion that wild salmon are still a major economic driver for tourism, recreation and industry.

From the sacred to the profane, the BC coast and those who live there are inextricably bound up with the destiny of wild Pacific salmon. For how can anyone speak of sustainability in our maritime communities without securing for the once-great runs at least a semblance of their former abundance?

Five species of Pacific salmon and the mighty steelhead trout run from the rivers and lakes, streams, sloughs and side channels of the vast marine ecosystem that connects the deep Interior mountains and plateaux of British Columbia to its wild coast. Some runs are already extinct, their spawning creeks buried in culverts under tonnes of urban asphalt or paved with gravel and debris from clear-cut mountainsides, their genetic stock exhausted by overfishing and lack of understanding of complex reproductive cycles.

Others, like the coho and steelhead that once teemed in streams on the east coast of Vancouver Island, now battle for survival against habitat loss, overfishing, climate upheavals and changing ocean conditions.

And now an exotic newcomer—the Atlantic salmon imported to stock fish farms—is on the anvil of an increasingly fierce debate. Its advocates say it's like a gift from God, a source of jobs for struggling coastal communities and of protein for a hungry world. Its detractors say it's a biological curse, the misbegotten product of a technological hubris that threatens the very existence of the wild Pacific salmon and the communities that rely upon it.

In the late summer of 2002, when word began to seep out that something dreadful had occurred in the Broughton Archipelago, where pink salmon runs were simply not showing up and starving bears were eating their own cubs, I decided to go and have a look for myself. That was when I began hearing charges that sea lice congregating around fish farms might be playing a crucial role in diminishing the ability of wild salmon smolts to survive. There seemed to be insufficient science to support the argument one way or the other. As I began fishing for answers, I found myself on a journey not unlike that of the salmon themselves, a journey that began on the spawning beds of little-known BC rivers, took me halfway around the world and brought me back, eventually, to the river where my own memories began.

The Broughton Archipelago

The silver-tipped massif of the Kingcome Mountains sweeps in a rugged arc around the still-pristine estuary where the lovely little Ahta River sings its way seaward to Bond Sound. This is among the last vestiges of the untamed salmon coast as it once existed from the Columbia River bar to the Alaska panhandle.

It was also, in the fall of 2002, at the eye of a political hurricane as British Columbia's $600-million salmon farming industry was accused of playing a crucial role in the disappearance of between 3.5 and five million wild salmon, triggering a fierce debate over whether economic development in the salmon farming industry is being achieved at the expense of wild stocks.

Once hailed as a new high-tech future, the industry is increasingly assailed by First Nations leaders for infringing on their aboriginal rights, questioned by scientists and economists, and the subject of a relentless battle among the spin doctors to win or deny public approval.

As the controversy over farmed fish and wild salmon deepened, I caught a float plane north to the remote and beautiful Broughton Archipelago, where I set out to track down the citizen-scientist who overcame intimidation by bureaucrats, indifference from government scientists and dismissals from industry to force the plight of wild salmon to the top of BC's political agenda. Yet to understand the magnitude of what Alexandra Morton has accomplished, I discover I have to turn first to the land itself to find a context that frames her work.

The Kingcome Mountains, outriders of the glacier-clad Coast Range farther inland, distill the fog, catch the rain and, far above the snowline, bank the frozen moisture from winter storms. When it melts each spring, the fast, clean, icy and oxygen-rich flow of water frothing down to the Ahta River below is what sustains the wild salmon and trout now stirring beneath its dark current.

Deep in the gloomy crevices of the bottom gravel, a new generation of pink salmon hatchlings has already begun to emerge from eggs deposited anywhere from five to eight months ago. For now, the alevins that have survived freezing, dessication, streambed scouring, hungry predators, the crush of stones and the threat of being washed out of the redds by downpour and flash flood remain in the safe haven between the pebbles, still feeding off their egg sacs.

Soon, usually at night, the emergent fry will fight their way free of the gravel and begin to rise into the water column for a feeding frenzy in which they gain the size and weight necessary to survive their journey to the sea. Then the green-backed splinters of silver will ride the current that gave them life toward the Pacific, where they will mature before returning thousands of kilometres to complete their two-year life cycle by spawning on the same patch of river bottom that nurtured them. Their outbound migration will, in turn, trigger another feeding frenzy as predatory insects, larger fish and waterfowl all congregate to feast on the hatch.

Though largely spurned by the sport angler and the commercial fishing fleet, the pink salmon is one of the most important components in a complex marine ecosystem that science increasingly recognizes as extending from the ocean deeps to mountain peaks. Pink salmon return in massive numbers on a two-year cycle. After they spawn, dead fish, some of which are dragged into the underbrush by bears or deposited as scat, will fertilize the bushes and trees of the riparian zone that shade and cool the water in summer and shower a steady sprinkle of insects upon the surface to be eaten by other salmon and trout fry, frogs, salamanders and birds. Other carcasses decay in the river itself, providing both a high-energy diet for larger mammals, eagles, ravens and seagulls, and a rich soup of nutrients for the insects and tiny organisms that will in turn feed the next generation of baby pinks when it emerges.

But how many of the pink salmon that usually teem in these waters will make their journey to the sea in the spring is a troubling question. And how

Salmon Farm Tenures in the Broughton Archipelago

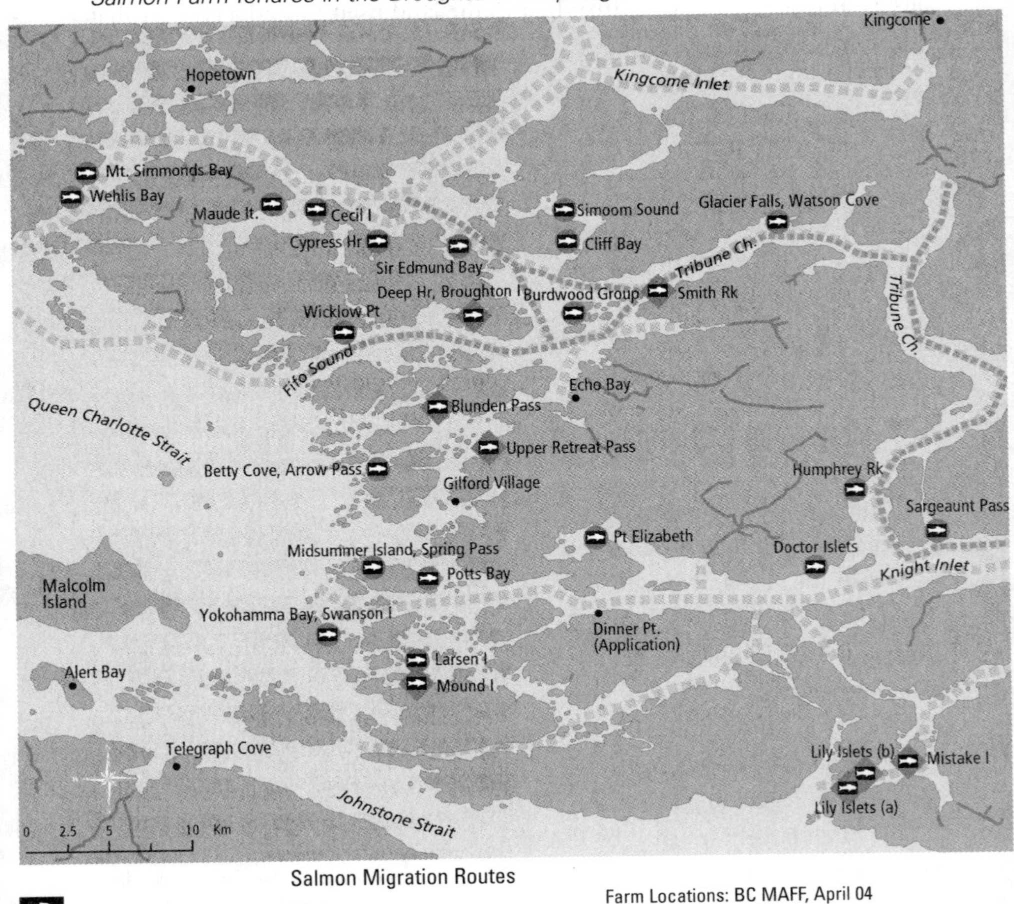

many of those will survive a gauntlet posed by one of the most intensive concentrations of fish farms on the coast is more troubling yet.

Veteran guide and local fishing lodge operator Chris Bennett fears it might be none. He has offered to run me out to the Ahta, the Viner and a few of the other creeks and rivers in the affected area. As we skitter across the cross-chop he tells me of his worry that in a collision between the artificial and the natural—between the industrial "progress" represented by salmon farms and the wild stocks that inhabit the region—a vast, complex and unimaginably rich web of interdependent life has already begun to unravel.

In the fall of 2002 the pinks that were supposed to lay down the eggs for the next generation failed to come sweeping in from the open Pacific. Their absence signalled one of the largest officially unexplained collapses of apparently healthy runs in BC history, a biological catastrophe of enormous proportions.

"Pink salmon runs have collapsed in virgin watersheds—what does that tell you?" he muses.

More than a few people, Bennett among them, are now pointing the finger directly at the salmon farmers. "I was here for a decade with no fish farms and I didn't see any of this," he says. "Now I've had a decade with the farms and I'm seeing all this stuff. It's not rocket science to figure out that there's a connection. It's going to go down as a very black period in the history of this coast. There has to be accountability in government for this kind of thing. Until there's accountability, we aren't going to solve anything.

"The normal run into this river is astounding. The whole river should be just black with fish. You could literally walk across the river on their backs," he says, gesturing to an estuary that's strangely devoid of the spawned-out carcasses that usually flush out with the freshets, attracting eagles and ravens. But in the fall of 2002, when the spawners known as the mainland inlet pinks should have been returning to the Ahta, Kakweiken, Ahnuhati, Glendale, Wakeman and Kingcome rivers by the millions, there were almost none.

Commercial fisherman Rick Burns, who has been fishing his own boat for 32 years, recalls the eerie silence that greeted him in late August 2002 as he turned his classic West Coast troller-gillnetter *Pacific Provider* into the remote, largely uninhabited Broughton Archipelago, a dense tangle of islands and channels between Vancouver Island and the mainland about 300 kilometres northwest of Vancouver.

Out in Queen Charlotte Strait, where he'd been fishing a sockeye run that surprised everyone, most of all the stock assessment experts at the DFO,

Burns had been catching 50 to 60 unwanted pinks a day, even with the much bigger mesh size that permitted the vast majority to swim to freedom.

"You saw pinks jumping everywhere," he says. "It didn't matter where you were. Johnstone Strait, Queen Charlotte Strait—what I can tell you about [that] summer was that it was a tremendous run for sockeye and pink salmon."

So Burns was expecting a bumper catch of pinks when he headed into Knight Inlet, where the previous year he'd hauled in 2,600 fish in a relatively short opening. The year before that he'd caught 2,000.

"In late August I went into the Broughton Archipelago and then ran 134 nautical miles into Knight Inlet and back, up one shoreline and down the other. It took me 24 hours and I never saw one fish jump. It was a watery wasteland. I knew that something dramatic had happened in Knight Inlet," he says.

Where he had expected to catch pink salmon by the thousands, he caught only twelve fish in more than 100 nautical miles of water, a number Burns says was verified by the independent observer put aboard by the federal government to make sure he abided by the stringent regulations requiring the mandatory release of stocks considered at risk like coho and steelhead.

The drama was not limited to the fish. Bennett says the big grizzlies that migrate to the river estuaries to fatten up on salmon before hibernation soon became desperate. Guides doing bear-watching tours on some of the key rivers reported that the grizzlies were gaunt and distraught and that in cases where there were two cubs, one of them appeared to have been killed and eaten by the others in the family unit.

"Without those pinks the bears were really hungry," Bennett says. "I saw one place where a bear had dug up an area the size of this boat looking for roots or insects. They normally do that kind of thing in the spring when they first come out of their dens. This I've never seen before." Bennett has been guiding in these particular waters for eighteen years, so that's saying something.

What caused the salmon runs in six Broughton Archipelago rivers to collapse will probably never be ascertained with absolute certainty, but both Bennett and Burns say they are convinced the culprit is the netpens of the twenty or so fish farms that stud the outward migration route of pink salmon smolts bound for the north Pacific.

Indeed, we've passed a number on the way down from Simoom Sound, where Alexandra Morton, a marine biologist who came to the area to study

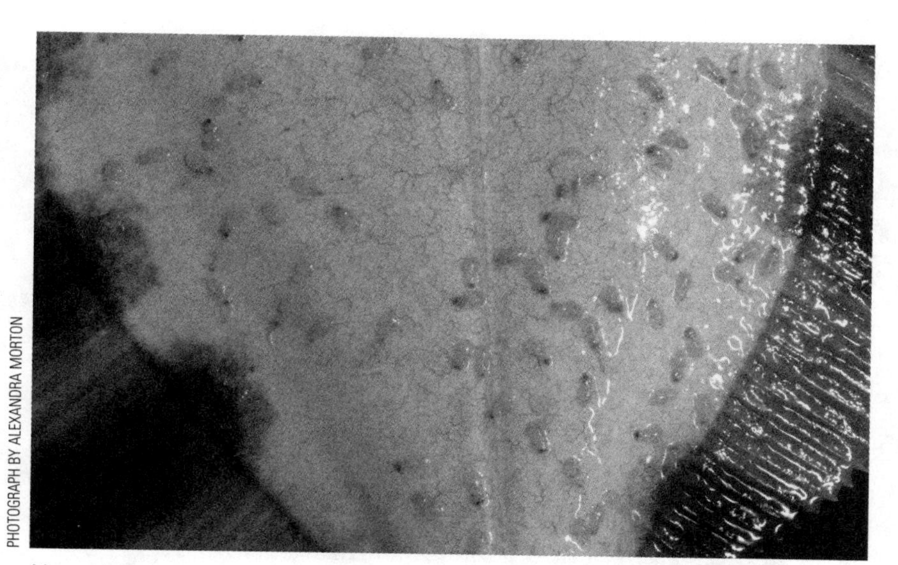

PHOTOGRAPH BY ALEXANDRA MORTON

Not only salmon get sea lice near salmon farms. This Starry Flounder was caught near a farm in Blunden Pass.

whales in 1984, first speculated that sea lice accumulating in vast numbers around the farmed salmon might have attacked the Broughton pinks.

The sea louse is a parasite that occurs naturally in BC waters, but wild fish are highly mobile and disperse over wide areas, minimizing the opportunities for the parasites to find hosts. However, "jam a million fish together like that and you have a sea lice ranch," Bennett says.

As the tiny smolts swam down the channels from their rivers, Morton theorized, they must have gone right through the schools of sea lice surrounding the netpens of the salmon farms. And because pink smolts are so small when they venture to sea—immature coho and chinook salmon remain in fresh water an extra year to gain the advantage of greater size—they would be far more vulnerable to the parasites than smolts from other wild salmon species. "I am absolutely positive that her complaints are legitimate," says Burns. "I guess the scary thing is that Alex Morton predicted this would happen back in 2000. And she was right. Now she's predicting extinction. I don't think we can afford to wait to see if she's right again."

"This sea lice thing is so clear. It's just common sense where they are coming from," says Bennett. "To talk about it is belabouring the obvious. If you see how the tide works in here, you know that the pinks coming out of Bond Sound go right along the side of the channel and they have to drift right past the fish farm pens."

When I meet Morton, she tells me that it was Bennett who first alerted her two years ago to the fact that something weird was going on with the migrating smolts.

"Chris Bennett had caught a four-centimetre-long chum and smaller pink smolt, both floundering off the end of his float," she says, leaning on the corner of a table in her home laboratory with its specimen jars and the computer with the satellite uplink that lets her exchange e-mails with other scientists around the world. "He was concerned because one of his guests [at the fishing lodge] had told him about what had happened to sea trout in Ireland." There, it turns out, sea lice predation on immature trout migrating to the sea had resulted in catastrophic collapses in stocks of the prized game fish.

Then she got similar information from Billy Proctor, an old-timer who can remember when the Ahta would choke with spawning fish, perhaps 60,000 seething in a pool two thirds the length of a football field and less than two metres deep, each female depositing more than a thousand eggs.

"I've never been out of the province," Proctor will tell you. "I've never had a driver's licence. I've never been to school a day of my life." But he has been fishing the waters of Bond Sound and the Broughton Archipelago for more than half a century, and lately he has been seeing things that make his eyes bug out—wild salmon smolts covered with sea lice, for example.

"That was awful to look at," he says. "I caught it with my ball cap. Any time you can catch a chinook smolt with your hat it tells you something. But it was floundering and dying."

Morton took her boat out to have a look. She had one of those epiphanies that only Mother Nature can deliver. She encountered the pink smolts migrating out of Bond Sound. "There were millions," she says. "There were fish everywhere, millions and millions of them. I didn't realize then that these vast ribbons of fish were the exceptional output from that big brood year in 2000." Morton began taking samples and discovered that as the smolts got closer to the fish farms, the prevalence of sea lice increased.

"Everywhere I went near the farms, the fish were covered with sea lice when I took them out of the water," she says. "Coho smolts were so frantic to escape the sea lice that they were jumping into boats. I noticed bleeding at their eyeballs and bleeding at the base of the fins, which are classic symptoms of fish disease. I was horrified to see these baby fish being ravaged by these parasites. It was an enormous feeling of helplessness." The scientist says she began to have nightmares in which she'd find sea lice attached to the eyeballs of her daughter, Clio, and would lurch awake in the middle of the night.

"Anybody who isn't moved by the grotesque image of those baby salmon being eaten alive by these sea lice—their little eyes popping out because the lice have eaten right through their heads . . . well, that person probably can't be moved," Bennett concurs.

Morton says she contacted a pink salmon specialist and asked if he'd ever seen sea lice on juveniles. He said no. She got in touch with a marine biologist in Norway. He warned her not to do the research by herself. So she took the early evidence to the DFO's Pacific Biological Station at Nanaimo and urged them to come and take their own samples. There she got only a limp implication that she'd spiked the samples and a request for more evidence, she says, so she went out in her small boat with a dip net and provided it.

Instead of a grateful thank-you from the government, a federal fisheries officer showed up at her door and informed her that she was under investigation for collecting samples without a scientific licence and could be liable for a fine of up to $500,000 and two years in prison.

Bennett becomes visibly angry just talking about the federal government's response to the initial warnings that a biological disaster might be unfolding. "It's scandalous, shameful how they've treated Alex," he says. "They've tried to discredit her, to marginalize her, but her facts have been bang-on. Alex is doing all this work because she loves the area. But she shouldn't have to do this. It's absolutely scandalous that she should have to do this and take the criticism she does. I mean, we have government people who are paid to do this research. Why is she left to do it by herself?"

Morton is more sanguine. "I had applied for a licence," she says, "but it takes three weeks to be approved and we had a very narrow window [before the migration was over] so I went ahead [and took samples]."

By then the whale biologist says she could sense which way the political wind seemed to be blowing within the federal and provincial agencies, so she got back on the internet and made contact with Rob Williams, a scientist working on similar problems at the Gatty Marine Lab, University of St. Andrews, in Scotland. She began studying sea lice intensively, something she'd never imagined doing in her wildest dreams. Between them, the two scientists designed a research project with appropriate parameters for spatial and temporal distribution of specimens and statistically valid sampling procedures. (Later, with three other scientists—one of them Simon Fraser University statistics expert Rick Routledge—Morton co-authored a research paper on the impact of sea lice infestations on migrating pink salmon in the Broughton Archipelago. It was published in the *Canadian Journal of Fisheries and Aquatic*

Sciences in spring 2004.)

"Sea lice are a copepod—a shelled parasite—and they have always been considered a benign parasite," Morton tells me. "Their reproduction rates are so low it's surprising they survive. But I counted 9,145 lice on 872 pink salmon smolts. I also found lice on pink, chum, coho, chinook, cutthroat and herring."

While a baby pink might be able to survive as the host of a single parasite, her specimens were averaging ten or more—and 78 percent of those she sampled were infected at or above the lethal level for a fish their size.

By now even the skeptics at DFO had to pay attention, and a team was finally dispatched to collect samples for the government. It produced an unsigned research paper that claimed few pink smolts had been collected. Scholarly critics outside government ridiculed the study's methodology. Environmental critics said it was evidence that the DFO did not want to find the same unpleasant data Morton had.

"They came to sample six weeks after I warned them and after the vast majority of the baby pinks had migrated away," Morton says. "They brought a dragger that's 50 metres long to try and sample fish that were a few centimetres. A dragnet smashes the fish against the toe of the net and all the lice are knocked off. Then they sent the seiner *Odysseus*. They caught only seven pinks in the area of concern. They did not sign names to the study. They did not submit it to peer review. Well, I can see why. It wouldn't stand up. You can't sample seven fish on the edges of your area of concern and say that it means anything. I seriously outfished them with a dipnet.

"I think it was public relations to defend the salmon farms," the biologist says. "I think they did it because they knew exactly what was going on. You know, first DFO said there were no lice. Then they said there was no crash [of the pink stocks] and that they had expected it. Then they modified that and said they had crashed everywhere. They said pink numbers were down significantly elsewhere, too. 'Significant' is a mathematical term. On the Phillips River the return was 50 percent of what was expected—but a 50 percent return is far different from a two percent return, which is what we appear to have here in these rivers."

Morton's findings set off an uproar among marine environmentalists and earned a spate of denials from the provincial government, which lifted a moratorium on the expansion of BC fish farms at the end of September 2002. Federal fisheries scientist Blair Holtby told a special meeting on the issue in Campbell River that neither natural variability in returns nor ocean conditions appeared to be responsible for the collapse of runs into the Broughton rivers, so inshore causes should be investigated. And the independent Pacific Fisheries Resource Conservation Council (PFRCC), headed by former federal fisheries minister John Fraser, subsequently concluded that there was indeed some evidence that sea lice could have caused the excessive mortalities among pink salmon smolts that resulted in the collapse of runs to the Ahta and five other rivers.

Nine months later, in February 2003, a research workshop in Vancouver convened scientists from around the world to discuss the apparent connection between sea lice and fish farms. Some urged more research. Others insisted that the precautionary principle required BC to act swiftly to mitigate an apparent problem. A few days later the province finally announced what it called a tough new control regime, which Morton dismissed as "too little, too late."

The government's plan sought to address concerns that the Broughton collapse was caused by sea lice from fish farms and to identify other inshore conditions that may have contributed. It combined two options proposed by the PFRCC the previous November, establishing a pink salmon migration corridor by strategic fallowing of eleven of 27 farms and accelerated harvesting to create a migration window. The government also imposed an enhanced monitoring regime for both wild and domestic salmon, requiring all farms to monitor and report lice numbers monthly (twice monthly during migration periods) as a condition of their operating licence, with results posted on the ministry's website.

While government and industry claimed that fallowing all but four of the salmon farms on the pink migration route was a positive step, Morton describes the provincial plan as "a brilliant deception. We have to fallow all those farms when the immature pinks come down. There should be no farm fish in the water during the migrations of wild fish."

"I would like to live long enough to hear just one fish farmer acknowledge that they might be causing a problem," Proctor adds. "They just blame it all on the wild fish."

Meanwhile, as the politicians tap dance around the issue, the scientists wrangle and the propaganda wars rage among various factions, pro and con, the vast, relentless and irreversible machinery of nature is already in motion, preparing to dispatch the remnants of the Ahta River run on their eternal journey to the sea. Or will this migration prove to be, as Alexandra Morton fears, their final journey to eternity?

Ireland

Out here at Ballynahinch, on the west coast of Ireland, I am in the remote, sparsely populated Gaeltacht where I can still hear Erse—the lovely, lilting Goidelic tongue that is the aboriginal language of Ireland—spoken on the street as casually as one might hear the musical cadence of Heiltsuk back home on the docks at Bella Bella.

Just as they do on British Columbia's mid-coast, the rain-laden weather systems scud in continuously from the open ocean, draping the steep hills with mist and depositing the precipitation that feeds the deep, cold lakes, or loughs as they call them here, and countless streams and rivers. Right now the rain is freezing, the thick, granular drops slicking the narrow black roads and spattering heavily against our Land Rover's flat windshield. In the high country off to the north there's already a dusting of white above the snowline.

It's a different ocean on a different scale on a different side of the planet, yet the culture, the climate and the landscape have much in common with the BC coast.

So does the region's recent environmental nightmare.

"The awful thing is about lessons not learned," Dr. Greg Forde sighed when I climbed the narrow stairs to his cluttered office on the Salmon Weir in the bustling regional capital of Galway and told him the story of the pink salmon crash in the Broughton Archipelago.

Forde is a big, rawboned, forthright man who once worked in the fish farming industry but now works for Ireland's western regional fisheries board. He's been struggling with a similar collapse of wild stocks that he says is also associated with salmon farms all along the Irish coast. "What we're hearing from Canada is that the sea lice are a factor where salmon farms are located on the salmon migration routes," Forde said. "It's all déjà vu. It's the most frustrating thing to hear what's happened here has now happened in BC."

Just up the road, his office hidden in an ancient stone hut that used to house the men who worked the now-closed salmon traps in the River Corrib, Seamus Hartigan, manager of the Galway salmon fishery, echoed Forde's sense of history repeating itself. "It happened in Norway for years and we didn't pay any attention. It's happened in Ireland and you are not paying attention. Do you want to learn by other people's mistakes or do you want to learn by your own mistakes?

"Norway had some of the best rivers in the world for the production of massive salmon—they are just gone," he said. "Why couldn't we learn from that? Why can't you learn from us? Is the BC government willing to make a place in the scheme of things for indigenous species?"

And he sent me off here, to Ballynahinch, one of the great fishing lodges of Ireland, where the catch of wild fish by sport anglers has now plunged by 90 percent or more. My guide through these twisting country roads to out-of-the-way places is Dr. Patrick Gargan, the senior research scientist with Ireland's central fisheries board in Dublin and one of the world's best-known experts on sea lice and the crisis in wild stocks that seems to occur wherever salmon farms locate along migration routes for juvenile fish.

While we drive, he explains how the life cycle of the sea trout resembles that of the pink salmon. Both leave their rivers soon after hatching—the sea trout as a survival strategy for escaping acid-rich, nutrient-poor drainage from peat bogs—and arrive at salt water in large numbers but of a small size relative to other fish. It's this size differential that makes the smolts more vulnerable to infestation by sea lice, especially when the parasites congregate around netpens containing millions of domesticated farm salmon.

"The smolts come out and they just get hammered," Gargan says. "They are constantly exposed to sea lice. They flee, those that survive, back into the fresh water to get rid of the lice. We find them in our sampling traps covered with lice."

It took some time to make the connection, he says. Even then it came only after a rigorous analysis had eliminated other possibilities. "We went through about fifteen probabilities. We looked at forestry, at climate change, at agricultural runoff, we even looked at the aftermath of radioactive contamination from Chernobyl. It all came down to one cause," he says. "Salmon farms."

At Gowla, Gargan clambers over a sheep stile, shoos away the curious woollybacks and leads me along the edge of a rocky outcrop. A jet of glassy, tea-coloured water plunges into a deep, tranquil pool. On the way in to this site, we'd passed the Zetland Hotel, once a big fishing lodge.

"There are no anglers there now," he tells me. "This fishery is closed since 1990. We have about 4,000 [sea trout] smolts going out and 50 or 70 fish coming back. It's difficult to keep selling yourself as a sea trout fishery when you are totally dependent on what they do to control sea lice out in the bay."

Gargan is here to show me a series of traps that monitor sea trout leaving and returning to the system. He's guardedly optimistic this time, predicting that there will be a significant increase in marine survival for the generation of immature fish bound for the sea this spring. That's because the salmon farmer operating in the bay where the river empties has sold out, and there are no mature Atlantic salmon in the netpens. Gargan says the new operator is just now stocking his pens with smolts, which means there will be fewer sea lice to attack passing sea trout juveniles. That's one adjustment he'd like to see for all the fish farms adjacent to migration routes for wild fish.

"What we want would be to have the smolts in close, because they'd not have the sea lice concentrations, and the big fish farther out," Gargan says, squinting into the wind and pointing out to the bay, just visible from the top of the outcrop. "But that doesn't work for the fish farmers. They want the big fish closer in because it's easier to harvest, and they want to harvest in August, not March. So it's all backwards. We want the farms to be fallowed in March."

Gargan has made this argument for sea lice management before, and in British Columbia, no less. He presented a paper at Simon Fraser University in July 2000 that warned of the lessons to be learned from the impact that sea lice had on populations of wild salmon and sea trout in Ireland and other European countries. Based on data gathered in 52 river systems around Ireland over eight years, his paper correlates the growth of the salmon farming industry along the country's west coast with rising infestations of sea trout with sea lice. It analyzes the lice infestation levels in rivers close to and distant from salmon farming areas.

He also offered advice in the paper on how to manage salmon farms in ways that would reduce or prevent the adverse impact of sea lice on migrating juveniles from wild stock. Gargan said that the Irish experience has shown the most effective way of reducing lice infestation pressures and increasing marine survival for wild stocks is to operate whole bay fallowing in the spring. When it's not possible to ensure that netpens are empty during smolt migrations, it is critical to ensure that lice levels are as close to zero as possible when the wild fish migrate, Gargan said. With proper lice management on salmon farms it's possible for depleted wild stocks to recover.

"You could have the salmon farming if you put your fish on a ten-month cycle," he tells me. "Or you could say no salmon farms in the estuaries of important wild fisheries. Or the salmon farms could be fallowed in January and left until March. Or you could have zero tolerance for sea lice in the cages. But it's like saying to an alcoholic there'll be no problem if you'll just not drink."

There are plenty of places in Ireland where the sea trout are doing very well, it turns out, but there are no salmon farms and their attendant sea lice anywhere near those river systems. The David Suzuki Foundation reported similar results when it sponsored a study sampling wild juvenile salmon in

BC's north coast waters, where there are still no fish farms. "A Baseline Report of the Incidence of Sea Lice on Juvenile Salmonids on British Columbia's North Coast," produced early in 2003, was the first in the province to collect data on sea lice conditions in pristine habitats before the arrival of salmon farms. That study found an average of 0.01 sea lice per fish—or 99.9 percent fewer sea lice per fish than Alexandra Morton found when she sampled pink smolts near fish farms in the Broughton Archipelago.

Salmon farms came to the Irish coast about twenty years ago for the same reason they came to the BC coast—clean waters sheltered from storm surges, privacy and ideal rearing habitat. Powerful tides flush through channels between a maze of islands and seethe against a rocky shoreline broken by deep fiords and backed by rugged mountains. Storms sweep in from the grey North Atlantic to deposit their rain into countless watersheds and wetlands.

The primeval forest is absent, of course, and counties like Mayo, Galway and Clare are characterized by bleak, wind-blasted moors, peat bogs and austere hillsides, yet, like the BC coast, this cool, rain-soaked landscape with its countless lakes and a vast network of swift-moving creeks and rivers provides ideal habitat for migratory fish, the emperor of which is the wild Atlantic salmon. And just as BC's prized sea-run rainbow, the steelhead, is the prince of fish for sport anglers, in Ireland there is the silvery, deep-shouldered sea trout, a migratory member of the brown trout family.

Since the days of Izaak Walton, elite anglers have made the pilgrimage here to test their skills with the fly against the sea trout. But this beautiful, untamed coast of western Ireland shares another, more sinister parallel with BC. Almost everywhere that a salmon farm has been located on a migration route for sea trout smolts, that local stock has collapsed in the same way that pink salmon stocks have collapsed in the Broughton Archipelago.

"Salmon farms get to a critical mass and then the lice from those farms concentrate around them," Greg Forde had said before I set out for the wild back country. "We have photographs of baby fish bleeding to death." He pulled out a file folder and showed me a sea trout smolt covered with sea lice. It looked identical to the specimens collected by Alexandra Morton back in BC.

"I'm a biologist," he said. "I have no doubt that sea lice from salmon farms are a major factor in mortality for wild salmonids. [In Ireland] you're talking about a catastrophic population collapse—sea trout runs of 20,000 going down to hundreds. We've lost a wild sea trout angling fishery that was worth millions of pounds. Fisheries here that used to be phenomenal are now derelict."

Out here at Ballynahinch, where a stunningly beautiful river winds past the window, Patrick O'Flaherty manages the 250-year-old fishing lodge for an American syndicate. Twenty years ago it was one of the most renowned in Europe. "What can I say?" he tells me. "This has been one of the most famous sport fisheries in the British Empire for more than 100 years. Tourism angling in an area like this is the very top end of the market. The average spending for a guy who caught one fish would be US$570 per day, and that didn't include anything except what he spent at my hotel.

"Now we're witnessing the death of the fishery. We've had to spend millions in order to try and secure diversified markets—cyclists, walkers and birdwatchers and the like—if we're to survive," O'Flaherty says. "The impact locally is heartbreaking. There were somewhere on the order of 25 boats out on Ballynahinch; each of those boats represented a ghilly [guide]."

Everywhere I stopped to talk to people about fish, I got the same story. At Glenicmorrin, fisheries manager Terry Gallagher said that since salmon farms located in the bay he'd seen runs of 10,000 sea trout dwindle to 50 fish. "This place has some of the greatest sea trout fishing anywhere in Ireland," he told me. "Everybody's known that for so long. Putting a [salmon] cage anywhere near this place is simply madness. But they've moved cages closer to the river mouth, so we're right into it now."

At Screeb, retired fisheries inspector Paddy O'Flaherty, who guided in his youth—his father and grandfather had been ghillies on the lakes and rivers overlooked by his house—told me that five years after fish farms located in the area, the sea trout population collapsed.

In his father's day, he said, the sport fishing kept seven men in work through the year. "Now there's nothing. There wasn't even 30 sea trout last year. The ones that were caught were in very poor condition, thin and covered with sea lice, blood coming out through their heads."

After we leave Ballynahinch, Gargan and I stop at Killary Bay to watch a helicopter lift plastic totes full of Atlantic salmon smolts from a gigantic eighteen-wheeler and deposit them into farm netpens out in the bay. Gargan tells me the Delphi River, which empties into this deep inlet, is the first river where sea lice-infested smolts were found.

Deep in the mountains, at the house on the Delphi that Lord Sligo built in the 1830s, I track down Peter Mantle, who left his job with the *Financial Times* in London almost seventeen years earlier to fulfill his dream of owning a fishing lodge and instead found himself battling to stay solvent. "This is where it all started," he tells me rather ruefully. "It's not been the romantic idyll we thought it was going to be. This house was built for sea trout. That's why it's here. Almost overnight we became a fishing lodge with no fish—a pub with no beer."

In May 1989 his fisheries manager reported a school of sick-looking sea trout smolts. "There were hundreds and hundreds," Mantle recalls. "The better part of our smolt run was in the bottom two pools of the river. We netted about 60. They were infested with juvenile sea lice. I rang every fisheries scientist I knew. Dr. Oliver Tully's first question was: 'Are you anywhere near a salmon farm?' I said yes, there's one about six miles away."

By 1993, he says, "it was perfectly obvious that this was a sea lice–salmon farm problem," but government and industry preferred not to hear. "The salmon farmers denied and denied and denied. Every time they came up with a counter theory, we had to respond and then they would rebut with what I have to say was crooked science. I hope the industry self-destructs—the sooner the better."

Mantle wound up working with a government task force seeking a cause for the sea trout collapses devastating the sport fishery. His recollection of that experience offers an eerie parallel to complaints from some scientists and environmentalists about the slow and ineffectual nature of the official response to collapsing pink runs in BC.

"They huffed and they puffed and said the evidence wasn't there until 1996, when even they had to acknowledge that we were right," Mantle says. "That group unanimously agreed that the problem that needed dealing with was sea lice and that until the sea lice level on salmon farms was reduced to negligible levels, the sea trout crisis would continue."

He's kept the Delphi Lodge going by opening to non-angling visitors and by starting a small hatchery, which uses the river's own genetic stock to raise and release wild Atlantic salmon smolts in the hope that enough will survive and return to provide angling opportunities. So far that's worked, although he's not entirely happy with the arrangement. "The extraordinary thing is how naive we were," he says. "We believed that once we'd identified the problem, steps would be taken to solve the problem. To this day I don't believe there is a single salmon farm that's been sanctioned for their chronic inability to control their sea lice infestations."

The Irish government, however, has brought in a more rigorous hygiene program and strict sea lice controls for the fish farms, which it claims are the most stringent in the world. Monitoring of sea lice levels occurs every two weeks during the smolt migration period from March to May and once a month the rest of the year. If levels rise above an average of 0.5 per farm fish, disinfection must take place.

But Greg Forde thinks there's an even better approach. At the very least, fallow all fish farms well before vulnerable smolts begin their migration. Better yet, leave major routes free. "There are areas that must be left virgin for the wild fish. Where we have bays and rivers with no fish farms, the sea trout runs are basically healthy," Forde says. "What we would recommend is that fish farms be located only in areas where there are no rivers with migratory salmonid stocks."

It's an argument that I hear again and again in BC. But the government's response has largely been to play down concerns and to respond with policies that resemble those in Ireland—more monitoring, more studies, chemical prophylaxis where necessary and a program encouraging expansion of the industry.

The First Nations

Oral histories collected by the anthropologist T.B. McIlwraith on the BC coast more than 70 years ago recorded bitter fighting over resources, the control of trade routes and complex dynastic politics. War over access to salmon rivers was a recurrent fact.

And today, as the provincial and federal governments push the promise of fish farming as a job-engine for small, impoverished aboriginal communities where wild stocks are in steep decline, the conflicts over resource jurisdictions and competing visions of a new economy are proving as deeply divisive as they ever were.

A few coastal First Nations have embraced aquaculture as an inevitable development. At last count, nine communities had projects underway, most of them shellfish operations employing long-established technologies that are generally considered benign by environmentalists. Only three communities were looking to finfish operations based on technology employing the more controversial ocean-based netpens for rearing salmon.

In response to these three, many former enemies from the mid-coast and the British Columbia Interior have become allies in a formidable opposition that formally rejects fish farms and other commercial enterprises in waters that lie within traditional territories.

In March 2004 the Union of British Columbia Indian Chiefs, a province-wide organization dedicated to securing aboriginal rights, wrote to the federal government to reiterate its policy of zero tolerance for sea-based fish farming using current methods. Founded more than 30 years ago, the group represents many Interior bands that have traditionally relied on wild salmon runs to the vast watersheds of the Fraser, Columbia and Skeena rivers and their tributaries. It warned that expansion of fish farm operations threatened the wild fishery and other marine ecosystems, citing findings by the Pacific Fisheries Resource Conservation Council and the alleged impacts of sea lice associated with penned Atlantic salmon on runs of wild pinks in the Broughton Archipelago.

The crucible for this emerging alliance is Bella Bella, an outer coast community of 1,200 on Campbell Island, about 180 kilometres north of Vancouver Island. Also known as Waglisla to the descendants of the six tribal groups of the Heiltsuk, Bella Bella is the centre of a territory that has been occupied by the same people for an estimated 10,000 years. Wild salmon are fundamental to the cultural, spiritual and economic identities of coastal societies like this one, and any perceived risk to the runs on which they have relied since time immemorial is seen as a direct threat to the survival not just of a village, but of the whole aboriginal culture. "We are struggling to save a way of life," Heiltsuk hereditary chief Edwin Newman told me.

Not everyone agrees, of course. The Kitasoo at Klemtu, a bit farther north, have hitched their wagon to the star of the fish farming industry. This community of about 350 has historically experienced unemployment in the range of 85 percent or worse. The reasoning behind its decision is based on a growing belief that the wild stocks which once sustained the community can never be returned to their former abundance.

The Kitasoo decision hasn't been without controversy of its own. Viral disease outbreaks have devastated their marine livestock in recent years, and they have found themselves marginalized in their own tribal community on the fish farm issue.

"It's hard to say no to the job opportunities," I was told by Harvey Humchitt, another Heiltsuk hereditary chief at Bella Bella, "but the risks [to wild salmon] are just so great that we have to do it."

"I shake when I think about our relatives from Klemtu, how they've been drawn into it, dividing us," says fellow chief Gary Housty. "We've got unemployment too, but we're not going to pay the price of being fish farmers."

This reluctance to view fish solely as an economic commodity suggests that the deep cultural importance of wild salmon to most First Nations eludes provincial authorities, small town mayors and many business leaders. It may be why perceived threats to stocks have emerged as such a deeply emotive

Netcages at the Klemtu salmon farm.

issue for so many coastal communities from Alert Bay to Prince Rupert—a region where aboriginals comprise 85 percent of the population yet remain seriously under-represented in mainstream political institutions.

When a delegation from the once-moribund Native Brotherhood of BC—membership had dwindled by 2001 to a mere 38 people—travelled the coast in 2002 aboard a vessel aptly named *Pacific Warwind,* it recruited thousands of new members from over 50 bands and First Nations between Old Masset in Haida Gwaii and Cape Mudge near Campbell River. Yet the divisions persist. At Kyuquot, a First Nations village on the west side of Vancouver Island that has no road links to the outside world, some members of the community agreed to salmon aquaculture in their traditional waters, but Leo Jack, who operates the Kyuquot water taxi service, told me he and a significant number of other locals remain opposed to any expansion by the fish farm industry.

"The only ones who don't go against the fish farms are the few who work for them. And that's not many," he said. "We had a meeting with the fish farm not long ago. They asked if they could expand. I asked how many fish farms they have. They said they have 24, all over the world. I said, 'How greedy does this guy have to get? How much money does this guy need?' This community needs to start saying no because these guys will never have enough. They will always want more and more and more until they have our whole inlet."

I heard this kind of rhetoric almost everywhere in the Native community. And although both the provincial government and the salmon farming industry seem unconcerned by what has been ignited—in the eyes of First Nations leaders, at least—by the unilateral decision supporting a rapid expansion of netpen aquaculture, the term "war" is now used angrily.

The fact that hereditary chiefs and elders have associated themselves with talk like this is particularly significant because it marks the political re-emergence of an ancient and long-suppressed government. These leaders bring with them a moral authority that no elected politician or corporate CEO from mainstream society could hope to command. So, when elders use militant language of a kind that hasn't been heard for more than a century in the troubled relationship between aboriginal people and provincial politicians, neither industry nor the BC government would be wise to take it lightly or seek to dismiss it, as some do, as the work of environmental agitators.

That assertion—blaming the heated conflict on outsiders—is seen by First Nations as the crude colonial assumption that aboriginal people can't be responsible for their own political activism, that they must be the pliable dupes of smarter non-Native agents. But there's no doubt that Chris Cook, the articulate president of the resurgent Native Brotherhood, is entirely his own man on fish farms or any other issue.

"When I look at farm fish, I talk about biochemical warfare," Cook said in December 2002 while addressing concerns about disease outbreaks, parasite infestations, unsanitary disposal of the dead farmed salmon (morts) and the potential for escaped Atlantic salmon to invade the habitat niches left by depressed wild Pacific stocks.

I listened to Cook state his passionate case at Alert Bay, a Kwakwaka'wakw fishing village on Cormorant Island about 350 kilometres northwest of Vancouver. He had been invited to an extraordinary meeting of the Mamaleleqala–QweqwaSot'Enox, 'Namgis, Da'naxda'xw Awaetlala, Tsawataineuk, Kwicksutaineuk Ah-kwa-mis, Gwawaenuk and Tlatlaskiwala—tribal groups within the Kwakwaka'wakw nation—convened by the Kwakiutl Territorial Fisheries Commission.

Delegates sat in silence as they heard about repeated disease outbreaks involving millions of salmon in fish farms around the Broughton Archipelago to the east and on the west coast of Vancouver Island. Cook didn't seem inclined to give the industry much quarter.

"I believe it's war on farmed fish," he said. "They are talking about 150 sites going north and each one of those sites will be where our salmon are going to be. We've been let down. Not just the Indians—more of the white people—we've all been let down in this province. It's time Native people came together with one voice."

Edwin Newman, invited from Bella Bella to speak at the December meeting, said that although salmon farming companies had attempted to divide and conquer by co-opting some poor communities with a promise of jobs, it was a doomed strategy. "We will not allow any fish farms in our territory," he said. "We will stop it any way we can. Even if it means going to war with our own people. The fish farm industry has been in Port Hardy for many years, but you don't see a prosperous community. You see a dying community. We will not allow our territory to become a garbage dump for the fish farm

PHOTOGRAPH BY GREG HIGGS, FOREST ACTION NETWORK

Youth protesting the construction of an aquaculture hatchery in Ocean Falls, January 2003.

industry. We are prepared to pay the consequences if we are to survive as a people. We have got a hell of a fight [ahead of us]. If we are going to stop them, we have got to do it together."

Greg Wadhams, a leader of the 'Namgis people who are centred on the Nimpkish River, with its once-vast runs of wild salmon, echoed those sentiments. "We are going to have to go to war with these guys," he said. "The only thing the government understands is what went on at Ocean Falls. It's about corporations and investors making money out of this resource and the rest of us starving."

What went on at Ocean Falls has become a lightning rod for this growing aboriginal rebellion. It's where the Heiltsuk village known as Laig was removed about a century ago to clear the way for a pulp mill. If this occurred someplace like Kosovo or Zimbabwe, we might call it ethnic cleansing. Now the forest industry has moved on, leaving behind only decaying buildings, scabbed concrete and the dank stench of abandonment.

Yet despite the original inhabitants' long-standing claim to the site, the province, after lifting a moratorium on fish farm expansions in late 2002, unilaterally approved a massive hatchery and smolt-rearing facility for Omega

Salmon Group, one of the multi-national salmon farming companies that contributed to the election campaign of Stan Hagen, BC's sustainable resources minister at the time. (Omega is a subsidiary of Pan Fish of Norway, the world's second-largest aquaculture firm.)

Approving such a facility in the middle of a hereditary chief's territory, territory subject to a specific land claim, sent a symbolic message that insulted elders in a culture where insults are justification for serious retaliation. Not surprisingly, Ocean Falls became the site of robust protests by the Heiltsuk from Bella Bella and the Nuxalk from Bella Coola.

I went out to see what was happening at one such protest in mid-January, 2003. I climbed out of the float plane early in the morning, then watched it take off and dwindle into the distance. An Arctic outflow poured frigid air into the inlet from Interior glaciers, and the dock was almost deserted except for the big patrol vessel *Inkster,* its tinted glass windows hiding the RCMP observers. The huge white hull dwarfed the few trollers and gillnetters.

I went for a stroll through what's left of the townsite, but hadn't gone far before one of the non-Native inhabitants, his face contorted with anger, snarled at me. "You're a visitor to our community who is not welcome," he barked and swore at me several times to make sure I got the message.

Bill Robson took pity on me after this performance and invited me aboard his fish boat, *Dean Kingfisher,* to warm up. Jammed in beside me was the Nuxalk hereditary chief Snux'y al twa—Deric Snow to the rest of us. He and Robson had come around from Bella Coola the day before and were awaiting the official delegation from Bella Bella, still hours away by boat.

"The government is going to have a tough time with this one," Snow told me. "What they are trying to do is divide the people and confuse us. It's not going to work. We are going to stand together. We are going to stop these fish farms. That's one thing we know, we don't want these fish farms."

Later I watched Heiltsuk hereditary chiefs come in on rafted fish boats. They were greeted by a welcome song from the Nuxalk chiefs in their regalia of headdresses, crimson and navy button blankets, and ceremonial aprons decorated with coppers and shells. There was even a Lummi chief, who arrived by float plane from the balmy Strait of Georgia far to the south.

These dignitaries then led a walk of close to 200—probably twice the remaining population of Ocean Falls—from the dock to the disputed site. At the gate they were stopped by RCMP officers in weapons and body armour. Nobody wearing a mask could proceed.

As it was, there was one mask—not some bandit's bandana, but the intricately carved Man-from-the-Sea mask that's worn for traditional winter dances—adorning a chief's nineteen-year-old daughter. The symbolic confrontation between the stupid power of state bureaucracy and the subtle moral authority of far more ancient traditions briefly took on the air of a Monty Python skit.

The RCMP officers, perhaps realizing how foolish they appeared, relented slightly. The young woman in the Man-from-the-Sea mask could pass, but only if a chief vouched for her.

"We told the RCMP the day before that this was going to be a peaceful demonstration," Harvey Humchitt said later. "It was disrespectful of them. They had no reason to be there in full force, in battle fatigues—man, that's a silly response. I think it was disgraceful."

Gary Housty, the Heiltsuk chief Naci, whose territory the Ocean Falls hatchery occupies, pointed out that the elders gathered for the demonstration were doing more to protect wild salmon than was the federal government, whose fiduciary responsibility they were. "Why is our Canadian government so weak?" he asked.

I hitched a ride back to Bella Bella on Humchitt's boat, the *Clea Rose,* standing on the stern while the dark water slid by. My hood was up against the wind and I ate hot stew from a Styrofoam cup, warmed my fingers around a mug of black, aromatic coffee.

Night deepened, a line snapped and the tarp we'd rigged for shelter blew away. But I stayed out in the weather, listening to the hiss of rain on the sea's face, to basketball stories, to a chief suddenly pointing through the spindrift to a clear-cut, pale against the hillside, and saying, "That is a burial site. My father's father is buried there. These are the things that hurt us."

The BC government promised a new relationship with First Nations in its throne speech, but as far as the Heiltsuk were concerned, Victoria was still treating them like colonial peons. They sought an injunction blocking the Ocean Falls development on grounds the province and Omega Salmon Group Ltd. began the project without the proper consultation required by law.

Before I left Bella Bella that cold day in January 2003, Housty told me, "This experience we're going through has made us stronger. We're not stopping. We are going to become more aggressive. It's not about us anymore. It's about our grandchildren and their grandchildren. My last word is this: Omega is not going to happen. Their dream is our nightmare."

The last word, however, appeared to be with the courts. In September 2003, Justice Laura Gerow affirmed the duty of the Crown and other parties to consult First Nations where there was a possibility their aboriginal rights might be infringed. But she ruled that in this case the Heiltsuk hadn't made an adequate case that their rights might be infringed. She said that proper consultation had occurred and observed that the Heiltsuk policy of zero tolerance for fish farms damaged the argument that consultation ever was desired.

And yet perhaps the last word won't be with the courts after all. Perhaps it will come from the lab.

Marine scientists convened at Alert Bay in January 2004 under the auspices of Simon Fraser University's Centre for Coastal Studies. They had gathered to review the preliminary findings of the previous year's research into interactions between sea lice and wild and farmed salmon stocks in the Broughton Archipelago. Rick Routledge, a statistics professor at the Centre for Coastal Studies, said that evidence connecting fish farms to the crisis may be compelling, but it remains circumstantial and incomplete. The scientists concluded that more research was urgently needed.

However the province's Ministry of Agriculture, Food and Fisheries announced that, based on the findings of the previous year's research, it would extend its requirements for sea lice management in the Broughton Archipelago to include the entire salmon farming industry on the BC coast.

Yet scientists remained divided on the contentious issue. Alexandra Morton's co-authored paper in the *Canadian Journal of Fisheries and Aquatic Sciences* in spring 2004 presented evidence that wild salmon sampled in proximity to fish farms were more heavily infested with parasitic sea lice than those sampled where there were no fish farms.

Scott McKinley, who held a senior research chair at the University of BC, disagreed. He said there was no evidence that wild stocks were in decline due to salmon farming. He suggested other factors, such as ocean temperature changes, might be at play.

Morton countered that similar sea lice infestations coincided with declines in wild stocks in Norway and that scientists there were expecting similar events on the Pacific Coast of North America. McKinley said the conditions in Europe were different from those in Canada.

As this academic tit-for-tat goes on in the foreground of scientific debate, the Heiltsuk of Bella Bella, the Kwakwaka'wakw of northern Vancouver Island and the Nuxalk of Bella Coola remain in the background like a gathering force of nature. They haven't gone away, they aren't going to go away and, after thousands of years of harvesting marine resources from the fecund mid-coast ecosystem, they know the meaning of patience—and when and how to fight, even if the weapons have changed.

"We own the whole of this country, every bit of it, and we ought to have something to say about it," Bob Anderson said for the Heiltsuk back in 1913. Even though Judge Gerow ruled against them on the Omega hatchery, she reaffirmed the fiduciary responsibility of government and industry to consult. Just over 90 years later, Anderson's descendants are making the same argument, but this time they aren't just singing it from war canoes; they are posting their message on an internet website, and their talk of confrontation and survival is being heard by a global audience.

The Alaskans

A wiry, wisecracking firebrand whose red hair flies like a Viking banner, Pete Knutson has been wresting his living from the cold, dangerous waters off the coast of Alaska for more than 30 years. He's worked for skippers who went to sea with no radar and no survival suits, had boats go down, been skewered by fish knives, had friends drown. He's also seen his financial fortunes rise and fall with the price of salmon, like the groaning hulls of small fishing craft that ride the swells rolling in through the narrow channels from Ketchiken to Conclusion Island.

Knutson doesn't fit the blue-collar stereotype of a fisherman. He went back to school and earned a PhD, teaches college classes in cultural ecology when he's not at sea, has a passionate commitment to his community and brings powerful analytical skills to the social and economic challenges facing the working-class compatriots whose company he never forsook.

And he says few things give him the sinking feeling that the prospect of industrial fish farms and their progeny invading the still-pristine waters off Alaska's southeast coast does. "It's really disturbing that there is an expansion of these fish farms in British Columbia," Knutson tells me. "Putting salmon farms up by Prince Rupert is a really bad idea. It's really going to heighten tensions."

That's a view shared by Doug Mecum, director of commercial fisheries for the Alaska fish and game department. "Moving [fish] farms to the north end [of BC] is problematic for us. This recent situation with sea lice has everybody spooked," Mecum told me in a telephone call to his office in Juneau. "We still don't believe they should have done it until they had their house in order."

So far, Alaska maintains a 1990 legislative ban on salmon farms of the kind that sprawl from Puget Sound, which has eight, to the north end of

Vancouver Island, which has scores. I say "so far" because the fish farming industry would dearly love access to Alaska waters. Knutson says the industry is "pushing really hard."

A bill that went before the state legislature in 2003 would have permitted fish farms for sablefish and halibut—provided they ensured the protection of fish and game resources and improved the economy, health and well-being of the citizens.

That makes Knutson splutter. "We coexist with the ecosystem," he says. "These farm guys come in and just blow the wild ecosystem apart."

The Alaska bill would still have prohibited ocean farming of either Pacific or Atlantic salmon species—a far cry from the controversial situation in BC. However, Alan Austerman, the state senator from Kodiak who sponsored the bill, resigned shortly afterward to take a new appointment as special advisor on fisheries policy to Governor Frank Murkowski, and the bill was consigned to the limbo of a committee.

Skeptics like Knutson and David Harsila, a gillnetter from Bristol Bay who was then also president of the Alaska Independent Fishermen's Marketing Association, worry that such developments are the thin edge of the wedge for fish farming in Alaska.

"They have penetrated various boards," says Harsila. "Some of our major processors are among the major distributors of farmed salmon. How can we direct our efforts to untangle their conflicted web of conglomerates, agencies and government? That's the challenge."

Still, in 2002 both state houses passed a joint resolution "respectfully" urging the BC government to reimpose the moratorium on salmon farm expansion until the industry could demonstrate its ability to manage such problems as escapes and diseases that might put Alaska's wild stocks at risk. State experts pointed to an epidemic of infectious hematopoietic necrosis, a viral fish disease that can result in mortality rates higher than 90 per cent among immature stocks and 40 per cent or more among mature salmon. Between 2001 and 2003 the epidemic affected twenty fish farms, many of them in the Broughton Archipelago—an outbreak even more serious than an earlier episode that infected thirteen fish farms during the early 1990s. The scope of these recurrent epidemics, say the Alaskans, is evidence the provincial government has failed to develop adequate controls and prophylaxis.

"To me this disease risk is a major concern," Mecum said. "I don't believe the Canadian government has done everything it has to do regarding regulatory control measures and quarantine." He said unchecked mortality and the spread of the virus without control measures mean there is the possibility it could mutate into a more virulent form and infect wild stocks.

Mecum cited a 2002 paper by University of Alberta scientist John Volpe, published in the *Alaska Fishery Research Bulletin,* which says that government monitoring of commercial catches indicates Canadian and BC authorities have consistently underestimated the escape of farmed fish that can mingle with wild stocks.

Knutson likens the combination of diseased farm fish and escapes to unleashing "smart bombs" carrying potentially lethal biological payloads upon the wild stocks.

"It's a silent killer, a plague, basically," Harsila agrees. "You have farmed fish which are vaccinated, but they escape and can carry disease into wild stocks that are not vaccinated. In Norway, they put their wild stocks into extinction."

One of the issues that most concerns the Alaskans is evidence from BC of what they call "multiple spawning events" by escaped Atlantic salmon in Vancouver Island rivers like the Tsitika, the Adam and the Eve. On this basis, they say, there is every reason to believe that the introduced species may spawn elsewhere and establish colonies. Of particular concern is the fact that Alaska has thousands of unnamed and virtually unmonitored streams into which escaped farmed stocks could penetrate.

DFO officials like Andrew Thomson, who heads up the federally and provincially funded Atlantic Salmon Watch Program, charged with monitoring escapes from fish farms, is able to counter with statistics that assert only 34 of the introduced fish escaped from netpens in 2003, a dramatic reduction from the 9,282 that escaped in 2002 and the 57,890 that escaped in 2001. Industry supporters like Mary Ellen Walling of the BC Salmon Farmers Association say the reductions in escape numbers are the result of better training and improved technology designed to comply with more stringent provincial regulations. Nevertheless, the same statistics also confirm that more than 1.4 million farmed salmon of both Atlantic and Pacific species have escaped into the wild since 1987. And Volpe has raised the point that nobody knows for sure how many Atlantics may have been caught by commercial or sport anglers and not reported.

It's on similarly cautious grounds that Harsila describes the BC government's decision to allow the expansion of fish farming as "very, very alarming."

"I think your government has been overwhelmed by corporate interests," he says. "Alaska is the next target. When I look at the farmed salmon industry I sort of boil it down to the structure of multinational conglomerates and how they can influence government. Judging from how the large companies have been able to achieve whatever they want in BC, we have to be very concerned about it."

Harsila's views echo those of Knutson and Mark Johansen, whose boat *Cora J.* is tied up just behind Knutson's *Loki* in Seattle. Both say they'd been catching mature Atlantic salmon—probably escapees from BC fish farms—up in Clarence Strait on the east side of Prince of Wales Island. Others have been caught as far away as the Bering Sea between Alaska and Siberia. Some have been caught in Alaska rivers. Stream surveys conducted in BC in 2003 found 36 adult Atlantics and one juvenile, Thomson told a reporter for the *Victoria Times Colonist* in the spring of 2004.

"How can you say you are concerned about the stocks in the Skeena or the Nass when you are bringing in these exotic species and letting them loose in the environment?" Knutson asks. "It's a massive biological invasion. It's almost as though the government of BC is just kissing its wild stocks goodbye. It's just so naked—and there's going to be huge blowback."

Johansen agrees. "It's like bringing rabbits to Australia. I'm worried about the spread of disease. It happened in Norway and Ireland and Scotland—I don't know why you'd think it wouldn't happen here. Maybe they are trying to get rid of the wild fish and the small boat fishermen. It's easier to manage three or four big guys than a few thousand little guys."

Knutson says that a grassroots backlash against farmed salmon already developing in European and American markets will serve only to tar the province in the eyes of an increasingly environmentally minded public. "You should be developing your extraordinary wild salmon stocks. Why do you want to put these pollution centres with exotic alien species right in the

middle of your pristine waters? Those runs in BC should be every bit as productive as the southeast Alaska runs—and BC has every advantage. It has productive watersheds, clean cold water, lots of feed, it's closer to markets. Instead BC is putting all its eggs in this one basket environmental groups have identified as being unsustainable without massive subsidization at the expense of ecosystems."

Indeed, Alaska seems a model to which BC might aspire. Since 1990 the state's wild fishery has produced almost twenty billion wild salmon at a landed value from the boat of more than US$4 billion, a figure that dwarfs BC's production of all salmon over the same period.

That is why Alaska's polite request for a renewed fish farm moratorium in BC comes wrapped around some hard intentions. State legislators also asked the United States federal government to consider the "numerous negative effects that farmed salmon from British Columbia have on the economy, environment and fishing industry of Alaska" when it was negotiating future trade agreements and Pacific Salmon Treaties with Canada. One resolution called for the US to restrict imports of farmed salmon.

So I went to interview Knutson aboard his venerable blue and white gillnetter, wintering in Seattle's Fishermen's Terminal. The salty-tongued former head of the Puget Sound Gillnetter's Association swung down the ladder to pull out samples of wild sockeye and chinook—boneless filets flash-frozen at -50°, whole fish intended for everything from a family dinner to a wedding buffet, smoked salmon, even custom canned sockeye—that he markets directly to customers who seek him out rather than go to the supermarket.

Selling off his boat like this is an act of civil disobedience. He tells me the fishermen who tie up at this working dock that dates from the beginning of the last century have been in a major spat with the buttoned-down bureaucrats of Seattle's port authority, which had plans to develop the waterfront, turn moorage over to pleasure yachts and to block the sale of fish products from the boats.

Yet even before I get out my notebook, the first car pulls up and the business gentry stroll down to the old boat to dicker over thick, de-boned slabs of Alaska spring salmon for a special dinner. Before we are done, the interview is interrupted half a dozen times. There is a well-dressed Hispanic woman looking for a sockeye filet, a middle-aged couple hoping for smoked salmon—he's sold out—and another wanting a whole fish for the barbecue. Knutson directs Mike in mechanic's overalls, who is looking for smoked Alaska black cod, to another boat.

"This is the future," he tells me. "The globalization guys think it's their 'pharmed' fish"—he spells it to make sure I note the pun, which refers to the pharmaceutical treatments required to control diseases, fungal outbreaks and parasites in the netpens—"but it's what I'm doing, selling my catch of wild fish direct to the consumer. You want to pay $18, I'll sell you a whole sockeye. You want to pay $6, I'll sell you a beautiful filet with all the pin bones pulled."

Johansen concurs. I find him sweeping drywall mud in long arcs across the wall of the kitchen he is renovating. "I have to find a way to sell more fish or I can do this for a living," he laughs. "Well, actually I do sell lots of fish, but I'm just about giving them away to the canneries."

He says cheap farmed fish from Chile was dumped into the US market in 2002 and depressed prices for wild fish from Alaska's abundant harvest. "The biggest crime was our coho price. At the start of the summer it was fifteen cents a pound and twenty cents a pound dressed—they wanted us to clean them for five cents a fish. It was beautiful fish. I stopped fishing in August because the price was so poor," Johansen says.

But he took advice from Knutson, who is an advocate of direct sales and who has been building a family network. Knutson fishes with his eldest son in the season while his wife, Hing, manages distribution back in Seattle and his youngest son handles deliveries. Because the operation is small, it's incredibly agile. He can have a fish out of the water, on ice and ready to ship in less than an hour, where big companies can take days. He sells to repeat customers from the dock, delivers fresh by air freight and sells at farmers markets.

"What's cool about this is that you are constantly interacting with your public," Knutson says. "Niche market producers like me are going to have our own websites which list what we have on board. You'll be able to find exactly what you want and phone your order. Believe me, this is a huge, high-value market."

Johansen, for example, came south with almost 7,000 kilograms of coho that would have earned him less than US$2,500 from the canneries. In Seattle, selling direct, he moved it for closer to US$45,000—enough to save his season.

"In BC you have a really rich, abundant ecosystem. You should be developing that, not fish farms," Knutson says, wrapping up his sales for the evening. "It's absurd. I mean"—he looks up and gestures with a slab of his boneless salmon—"it all comes to you, and you shove that aside?"

The Fish

Out in front of the ten-metre welded aluminum boat, the crests have begun to feather off the waves and spatter against the windscreen to the steady beat of wipers, the troughs have shortened and the incessant pounding of the choppy seas against the hull makes it impossible to take notes, let alone think.

We are into the lean winter months of 2000 but the scenery here at Phillips Arm more than compensates. Rugged, snow-capped mountains rise out of a sea that winds away into a bewildering maze of fiords, blind channels and passages between forested islands. There are Steller's sea lions on the rocky headlands and eagles soaring on the thermals.

"We're going through a southeast to northwest shift in the weather," explains Don Tillapaugh, who's taking me out to see Liard Aquaculture's fish farm sites on this remote inlet and at Hardwicke Island, somewhat closer to Brown's Bay, where my voyage began and where the fish are processed. "Right now we've got a southeast wind coming over a flood tide, that's why the waves are stacking up. This can be some of the ugliest water in BC. Tide rips."

Up here, where billions of tonnes of water jet into and out of the Strait of Georgia with every tide change, is where almost half the farmed fish

Atlantic salmon caught in a gillnet in BC waters.

production in BC originates. At last count there were 84 fish farm tenures in the region that includes Georgia Strait and Johnstone Strait, and they produced about 66 percent of the province's farmed salmon harvest. I was told by industry experts just before leaving that about 2.5 million kilograms of dressed farm salmon leave Vancouver Island for US and Asian markets every week.

It's a fiercely competitive market and the consumer demand is for Atlantic salmon, which partly explains why 77 percent of the fish raised in captivity in BC waters are Atlantics. Canada's share of the US market for fresh whole farmed Atlantics grew from 70 to 82 percent between 1996 and 1998, and its share for fresh Atlantic filets more than doubled, from six percent to fourteen percent. The US market is expected by some industry analysts to continue to expand by about $100 million per year.

Liard Aquaculture alone accounts for six percent of BC's production I'm told as we head north. Not bad for an industry that's less than two decades old. But if things are booming now, it wasn't always this way. Like many sectors of agriculture, salmon farming has been affected by boom and bust cycles. In the mid-1980s about half the province's 150 or so fish farms failed because of problems associated with poor siting of their netpens and the mortalities from algae blooms and disease outbreaks that followed. Then in 1989 a record harvest of wild salmon caused farmed salmon prices to crash by up to 40 percent.

"Basically, anybody who was in the business in 1985 went broke," Tillapaugh explains over the din. "We were raising pan-sized fish, but the market wanted four- to six-pound fish. In the early days, price was high enough at $6 to allow us to learn from our mistakes. But then the price crashed because of oversupply, the Canadian dollar strengthened from 72 cents to 90 cents US and inflation took off, all within one year." He shakes his head. "You have to have knowledge in order to survive. You have to improve health. You have to reduce escapes. You have to improve conversion efficiency."

Tillapaugh says advances have been made on all those fronts. Where it used to take two kilograms of feed to produce one kilogram of salmon, that ratio is now down to 1.3 kilograms of feed for a kilogram of salmon. Mortalities are down to less than three percent at Liard Aquaculture. And the industry is committed to reducing escapes, which continue to alarm critics.

How big is this problem? To date more than 1.4 million farmed salmon, almost half a million of them Atlantics, are known to have escaped into Pacific waters, and there are believed to be thousands more that were never detected or remain unreported. Specimens have now been captured as far afield as the Bering Sea. Critics fear that farmed Atlantics could open a disease vector into wild populations with no natural immunity and that drugs used to keep farm fish healthy might spawn antibiotic-resistant supergerms in the ocean. Some worry that escaped Atlantics might eventually displace weak coho and steelhead stocks.

Another concern in Europe and on the east coast of North America arises from evidence suggesting that some escaped farm salmon have mated with wild stocks, apparently reducing the ability of the wild fish to survive. In BC waters, of course, this is less a concern since most scientists consider it unlikely that Atlantic salmon would be able to breed successfully with Pacific salmon. However, some Atlantic-Pacific hybrids have been bred successfully under laboratory conditions. A few offspring even survived three years or more, although the overall survival rate was extremely low. It's thought more likely that individual escaped Atlantics might compete with Pacific salmon for access to spawning females, thereby preventing them from depositing fertilized eggs, which could be significant in a threatened or severely depressed population.

Experts from BC's fish farm industry—and the federal and provincial governments that regulate and promote it—dismiss the notion of Atlantic salmon establishing themselves in West Coast rivers as paranoid fantasy. They cite unsuccessful attempts in 1904 and 1934 to stock BC rivers with up to nine million Atlantic eggs and fingerlings. A 1997 report prepared for the province's Salmon Aquaculture Review referred to these experiments: "The potential establishment of Atlantic salmon populations in Pacific rivers and streams as a result of reproductive success and survival of progeny in a series of years is considered unlikely, based on the historical failure to establish the species in waters outside its indigenous range."

But the report didn't close the door on the possibility that a new anadromous species might colonize BC streams, noting that "biological and ecological factors which have contributed to past failures are largely speculative and opportunity to colonize may have improved in recent years." The study cited the relatively large number of adults that have escaped (federal statistics show that more than 1,000 Atlantics have now been found in 81 BC river systems). These bigger, stronger adults would have a higher probability of surviving to spawn compared to the primitive release of eggs and fry that characterized the initial attempts to establish the fish. Furthermore, critics say, salmon farms raising Atlantics in BC are close to many potential spawning areas in which the population status of wild salmon runs are known to have declined, reducing the capacity of native stocks to prey upon the fry of an introduced species.

"Although the risk of colonization may remain low, the opportunity for successful colonization has likely improved," the report said.

However, not everyone shares those concerns. Net-reared Atlantics that escape from pens where they are fed pellets every day simply don't adapt to wild food sources, says David Groves, who produces 300,000 Atlantic smolts for resale from his Crofton-area hatchery. "Most of them, I think, simply die at sea. The probability of a surviving reproductive fish is almost zero."

Andrew Thomson, who has monitored escapes for the last decade from his Atlantic Salmon Watch office at the Pacific Biological Station in Nanaimo, corroborates that assessment. Of more than 1,000 escaped Atlantics tested, he says, only seven percent had food in their stomachs.

But in 1999 a young doctoral candidate donned his wetsuit and made a shambles of those comforting assumptions. John Volpe had already proved that Atlantics would spawn in BC rivers if they had an opportunity. He'd demonstrated as much in a controlled experiment on the Little Qualicum River. That experiment was abruptly terminated when the federal government suddenly took back the contained spawning channel he'd borrowed on grounds the "risks of negative impacts" were unacceptable. So Volpe turned to field sampling. From Vancouver Island's Tsitika River he collected twelve immature Atlantics. "They were like little footballs, they were so fat," he said.

How might they survive when Atlantics introduced 80 years ago perished? Volpe suggests severely depressed Pacific salmon populations could have left a niche vacant for colonizing Atlantics. "River systems back then were considerably different from today. Then habitat was saturated with native competitors and there was no ecological room for a new player." On the Tsitika today, however, steelhead numbers are so low that only half their habitat is used, leaving the rest available to an invading competitor.

The scientist's Tsitika discovery shook the conventional wisdom to its foundations. American Frank Utter, giving evidence at hearings into fish farming in Washington state, said that Volpe's findings in the Tsitika seriously undermined the BC Environmental Assessment Office's conclusion that escaping Atlantic salmon posed minimal risk to wild stocks. "This conclusion was largely based on the fact there were no examples of this species becoming established as an anadromous species outside of its native range. Now that we have such an example—from the very region in which the risk of assessment has been conducted—it is clear that the basis for the EAO's conclusions in this regard no longer exists."

Meanwhile, Volpe had found a sexually mature pair of Atlantics in the very first pool he entered on Amor de Cosmos Creek, not far from the Tsitika. "The next year we went back and it was just full of juveniles. We counted 116 fish in a three kilometre stretch," he says. "That's a minimum of 116 and likely there are a lot more. There were two age classes." Age differences are particularly significant because it suggests Atlantic salmon have successfully spawned and reared in the stream.

There was interesting news in the lab, too. Volpe had an independent DNA analysis of his specimens done at UBC. It confirmed the fish collected in the Tsitika were Atlantics. Furthermore, scale ring analysis indicated they were not reared in captivity—they were wild. Farm fish, which have a plentiful food supply and scheduled feedings, show a regular pattern of growth rings. Feral fish have rings that differ in width, reflecting food supply fluctuations. Embarrassing as it might have been to federal and industry scientists, Volpe had just provided the first strong evidence of Atlantic salmon repro-

Atlantic salmon with ripe eggs found in Johnstone Strait.

ducing in a Pacific river.

Here in BC, official response to Volpe's findings was hostile. Representatives of the aquaculture industry publicly questioned his scientific credentials, dismissing him as a mere student—he was then a few months from obtaining his PhD—and at DFO the criticism of his research even went so far as to imply his scientific integrity was at issue.

But Dave Anderson, a Washington state politician who represented a constituency among the islands at the south end of the Strait of Georgia, said the issue deserves more serious study by government, and he introduced a bill in the state legislature to that effect. "They [Atlantics] can't possibly spread disease. They can't possibly survive in the wild. They can't possibly reproduce in freshwater rivers. They can't possibly do anything," Anderson laughed when I telephoned to ask about his bill. "My take is that if you put enough of them out there over a long enough time they're going to adapt and occupy a biological niche. Who knows what they will do or not do? But given our experience with exotic species elsewhere in the world, there's cause for caution."

Scientists in the Alaska state department of fish and game expressed similar concerns in a paper drafted to brief the state government in February 1999, arguing that escapes of mature Atlantic salmon into the already challenged Pacific Coast ecosystem represented an enormous potential threat to wild Pacific salmon. "Introductions of non-native species have frequently resulted in unexpected and often disastrous consequences resulting from competition, predation, crossbreeding, or the introduction of non-native species or parasites." The scientists fruitlessly called for BC to continue its moratorium on major fish farm expansion.

In 2002, after BC lifted the moratorium, Alaska released a white paper that described BC's "managed risk" policy with respect to netpen salmon farming as a failure. It acknowledged that closing the industry down entirely was "simply not an option," but instead advocated a zero-risk policy that would include revoking an operator's permits in the event of a disease or escapement failure. It called for the "branding" of caged salmon so they could be traced back to their source if they escaped and to make it easier to identify Atlantics that had successfully reproduced in the wild. It also recommended that only sterile females be raised as livestock so they would be incapable of reproducing should they escape.

A subsequent study released in 2004 by the Georgia Strait Alliance, a non-governmental group concerned about environmental issues, shared the Alaskans' misgivings. It gave the BC government a failing grade in eight out of ten categories when it came to regulating the salmon farming industry. It claimed the government was failing to protect wild salmon, failing to listen to coastal communities about their concerns and failing to listen to the objections of First Nations.

When 30,000 Atlantics escaped from a single farm near the north end of Vancouver Island in 1999, marine biologist Alexandra Morton was prompted to write to Dennis Streifel, the provincial fisheries minister at the time, about her concerns. In a pool full of spawning coho she had caught a specimen covered with pus-filled boils, symptomatic of a contagious bacterial disease called furunculosis. "This fish was fresh off a farm, in a stream, carrying an infectious bacterial disease, in an extremely contagious condition, schooling with wild coho," she wrote.

Then in February 2001, Canada's auditor general warned the federal government, which remains a strong advocate of marine aquaculture, that it was not fully meeting its legislative obligations to protect wild Pacific salmon stocks and their habitat from the effects of fish farming. Federal fisheries officials were not adequately monitoring salmon farms for effects on wild salmon stocks and habitat, there were significant gaps in research regarding salmon farming and its impact and potential impact on Pacific salmon, and government needed a formal plan for managing risks and for assessing the potential cumulative environmental effects of proposals for new sites should the decision be made to expand the industry.

Although the risk of transmitting diseases from farmed to wild stocks is considered small by scientists in both industry and government, it's a public relations problem that long-time aquaculture scientists like David Groves say is unnecessary. "These [escaped] Atlantics are an embarrassment that they are even out there," he told me. "They don't need to be out there. And the industry has had a pretty hard look at itself lately. Escapes are something that the industry will have to bear down on hard because we don't want them out there."

Odd Grydeland, a long-time aquaculture manager who has served as president of the BC Salmon Farmers Association agreed. "Atlantics escaping from farms—that's one place we and the environmental groups have the same goal. It doesn't do anybody any good. We don't want fish to escape," he told me when I stopped by his office to ask. And he pointed out that there is a powerful financial incentive to prevent escapes. "Every time we see a ten- or twelve-pound Atlantic getting away, it's a $50 bill swimming away."

Stricter protocols and better equipment are one answer. But as long as fish are raised in ocean netpens and not closed containment systems, they will continue to escape, critics say. And given the demographics of the industry, the fish that escape will likely be Atlantics.

Originally BC fish farmers raised primarily chinook in captivity. Today the production of Atlantics dwarfs that of native salmon species, for several reasons. First, Grydeland said, there were problems with domesticating Pacific salmon. In some cases farms were located in the wrong place and were susceptible to algae blooms. Pacific salmon proved less efficient at gaining market weight. They are more difficult to handle and subject to greater stress from handling, which reduces the quality of the flesh. And there were disease problems as farmers tried to figure out the proper population densities for netpens.

Second, there was an infusion of European capital into the struggling industry, and with it came a knowledge base that was considerably advanced. "There was a large influence from blue-eyed Vikings coming over here from Norway with a lot of experience about raising Atlantics. They knew Atlantics, they had a head start of about a dozen years with developing the technology. And Atlantics were already being produced successfully in Washington state."

There were two other main advantages. "Atlantics have a lower feed conversion rate and they tolerate higher densities so we can have more fish in the pen and have a much lower mortality rate," Grydeland said. "And when they do die, they perish very early in the production cycle. When Pacific salmon die, they do so at around four kilos. If you are going to lose some fish, you want your fish to die when they weigh a few grams."

At the Hardwicke Island facility it's easy to see why that might be a concern. The pellets used to feed the fish are loaded onto the feed barge 80 tonnes at a time, a tonne every 30 seconds. It's dispensed to the fish at a cost of $15,000 a day during peak feeding periods.

"Atlantic salmon will eat two pellets a minute when they're hungry," Tillapaugh points out. "Sixty cents of every dollar we spend goes to feed, so efficiency is very important to us." That makes simple economic sense. If you're going to lose fish—and at Liard they claim a 98.2 percent survival ratio—you want to lose them before they've consumed all that expensive feed.

Behind Tillapaugh, the netpens form a long line on either side of floating catwalks. Each pen encloses 13,000 cubic metres of sea water and tens of thousands of fish behind a series of nets designed to keep fish in and predators like seals or sea lions out.

Tillapaugh gestures back to where a worker uses a blower to direct feed pellets according to strict feeding schedules. As the pellets strike the dark water, it seethes with feeding salmon. "There's a million-and-a-half dollars in there," Tillapaugh says. "Every fish in that pen is worth $40. We absolutely don't want to lose any."

The process has been honed to the finest tolerances. Underwater cameras monitor the fish in each pen and transmit to all-weather television screens. Computers tell fish farm workers exactly how much feed should be distributed. Automatic feeders distribute it precisely. Sensors under the nets monitor how many feed pellets pass through the cage—if too many pellets are falling through, it means the fish aren't hungry and the feeder is shut down.

"There is a lot to feeding fish," Tillapaugh says. "Fish are cold-blooded, so if the temperature goes down you know they are going to go off their feed. If a storm moves in and the air pressure changes, they might go off their feed."

And then there are the logistics of keeping the fish healthy. "I don't know how many smolts there are industry wide—30 million or so," he says. "Every one of them gets a needle. All our fish are injected with a multi-valence vaccine."

This is another source of controversy—and not just with environmentalists. The Alaska government has expressed concern about the pharmaceuticals and vaccines required to permit high-density populations in netpens. In their white paper, scientists from Alaska's fish and game department warned that crowding the fish increases stress and makes the penned salmon more susceptible to disease. When disease outbreaks do occur they tend to spread rapidly through the captive population. For this reason many farmed salmon are inoculated, when possible, against disease agents. The Alaska scientists were concerned that "the unregulated prophylactic use of antibiotics may result in diseases mutated into more virulent or antibiotic resistant forms." Indeed, antibiotic-resistant strains of pathogenic bacteria are reported to have emerged on both Canadian and Chilean fish farms.

Still, even the Alaskans acknowledge that so far all pathogens reported in farmed Atlantic salmon have been indigenous to local wild salmon stocks and that the risk of farmed stocks infecting wild stocks is probably low.

Groves says that when BC producers began importing Atlantic salmon eggs to create a broodstock back in 1985, DFO scientists were "bloody-minded in their conservatism" when they set up hygiene protocols. As a result, he says, "we don't have anything here that wasn't here already. We have a healthy Atlantic salmon broodstock and we now don't have to bring anything in."

Groves says the industry has also become much more sophisticated at identifying low-level endemic diseases and preventing outbreaks. "It's like any other animal husbandry. You never relax in your vigilance against disease. You just never relax."

Meanwhile at the Liard Aquaculture site I visited, reliable as clockwork, the netpens kept on producing four truckloads of high-quality dressed salmon and filets for the world market every week. They get there aboard ships like the *Wespak,* so I followed along to observe the end of the production process.

On this day I watch as big Atlantic salmon swim idly in the chilled, heavily oxygenated water of a series of large holding tanks on the transport vessel. In a matter of minutes they'll be scooped out, water and all, dropped into a slurry of ice and water charged with carbon dioxide that stuns them, and sent down a conveyor belt that takes them into Brown's Bay Packing Company, just north of Campbell River, where they'll be bled, dressed, chilled and either packed or directed by the pallet-load to the value-added plant next door.

Inside, workers in hairnets and white lab coats work swiftly to clean and pack the fish. Through a set of heavy swinging doors, another team of workers runs salmon through an automated filleter that spits the backbone into an offal bin. Fins and other bones are removed, the skin is stripped by another machine and the two filets are placed together for packing—a chilled, perfectly fresh boneless salmon. At the end of this process, 60 percent of the fish carcass has been recovered as usable food. The heads are sold to a separate specialty market. The offal goes to a composting operation.

"The orders come to us, we pull the fish out of the cooler and pretty well cut to order," says the operation's general manager Ken Pike. "This seems to be the direction in which the industry as a whole is going—this is where our customers are telling us they want us to go."

The time from live tank to filleting table is measured in minutes, a process that keeps the product at the premium end of a picky but increasingly lucrative market.

"This is beautiful stuff," Pike says. "I spent fifteen years in the commercial fishing industry and I know quality. It's a really simple process. It depends on how long you can keep the salmon alive before slaughter. You want to minimize the time from mortality to pack."

He says the consistent and predictable supply of salmon from the fish farms scattered across the region is what gives British Columbia's growing aquaculture industry its competitive edge. "You have the ability, in theory, to

NET LOSS OF PROTEIN IN 1999

2,126,000 tons of fish taken from the oceans... ...to produce **871,200 tons of farmed salmon**

control the quality of the output," Pike says. "It's not like speculating. If you're at a wild fish plant, for example, you don't know whether you're going to have 100 pounds of fish at the dock in the morning or 100,000 pounds. We're market driven where the commercial wild side is volume driven—let's go catch everything we can and hope that we can sell it."

Brown's Bay Packing is an example of what supporters of the fish farming industry see as the future for coastal communities hit by declines in logging and the commercial wild fishery. The plant has been in the custom processing business since 1989, and the company's promotional literature says it injects more than $6 million a year in wages into the Campbell River economy, as well as an additional $3 million in the purchase of goods and services. Although the workforce naturally fluctuates according to market demand, when I visited the company employed as many as 200 people. On the day of my visit there were 34 at work in the primary fish plant, where they gut and dress the whole fish, and another 45 in the value-added plant, where salmon are either cut into steaks or processed into skinned, boneless filets for the US market or the fresh retail market in Calgary and Edmonton. The shifts are regular—the plant can process 45,000 kilograms of fish with two shifts—and the average wage is better than $17 an hour.

"Most of these people have been with us since this building opened," plant manager Ross Campbell explains. And if the US market for value-added salmon products continues to expand at its present rate, they're likely to be there for a long time yet. Certainly the federal government has that expectation.

At the end of 2003, federal aquaculture commissioner Yves Bastien brought down his annual report on the state of the industry. He observed that while fish farming is the fastest-growing food production sector worldwide—the industry in Canada grew at an annual rate of nineteen percent between 1986 and 2001—the forces driving its expansion are both immense and global in scope. World population is expected to expand by 36 percent between now and 2030, with a corresponding growth in the need for protein for human nutrition. It's a requirement that cannot be met by relying on traditional wild sources in any sustainable way. As a consequence, Bastien's report says, world aquaculture output is expected to surpass beef production by 2010 and will continue to show a high growth rate in Canada as well as elsewhere. While fish farming employed about 12,000 people across Canada in 2001, it has the potential to employ more than 47,000 by 2020 and to more than quadruple its farmgate value to $2.8 billion a year in the same period.

That may be good news for investors in the industry, but what will it mean for wild salmon stocks? Expert opinion remains deeply divided. One thing does seem indisputable. To achieve the kinds of agricultural outputs from aquaculture envisaged by the industry and its visionaries will likely require an industrial transformation of BC's coast similar in magnitude to the fencing of the prairies and the conversion of wild forests into fibre farms. For good or ill, the future of the BC coast is now inextricably entangled with Atlantic salmon and their fate.

The Farmers

As I made my way north on the Island Highway to Campbell River, the self-styled Salmon Capital of the World, stars glittered in the clear, black sky like a million chips of ice. By the time I was in my hotel room, though, looking over Discovery Passage to Quathiaski Cove and Quadra Island, the clouds had settled thick as a Cowichan sweater. A stiffening southeaster rattled my windowpanes and spattered the glass with pea-sized drops of rain.

It might have been a metaphor for British Columbia's salmon farming industry, which began in a sunny mood of optimism and clarity of purpose, but now finds itself under a murky cloud of controversy. Lashed by critics that some fish farmers denounce as liars and rabble-rousers, the industry is dismayed by what it sees as a one-sided response from the media. It seems stunned by the intensity of the public debate that erupted following the catastrophic collapse of 2002's pink salmon runs to six rivers in the Broughton Archipelago.

But Mary Ellen Walling, the executive director of the BC Salmon Farmers Association (BCSFA), who came from a career of development banking and community-building with North Island College, tells me she is determined not to let the controversy slide into the polarized trench warfare that characterized the province's forest debate in the 1990s. So perhaps it's no surprise that an association website urging members to help counter media negativity by writing letters to the editor—politicians consider the ideas expressed to them in a letter to be representative of up to 1,000 voters, it optimistically advises—also urges diplomacy and courtesy.

Walling believes both salmon farms and environmentalists are here to stay, which means that both sides, if they are motivated by genuine concern, will have to grope their way toward that difficult middle ground where policy can be forged from consensus rather than rhetoric. She is also quick to acknowledge the dominant place the wild salmon holds as a totemic figure at the heart of British Columbia's collective sense of cultural identity, an icon whose importance unites First Nations and non-Natives alike. Indeed, when DFO conducted a series of consultations across the province in 2000 as it was crafting its new wild salmon policy, it heard so frequently that it was essential for the federal government to ensure the conservation of stocks that it concluded: "Wild salmon are central to the culture, the ecosystems and the economy of BC."

In a Sierra Club report on wild salmon, Terry Glavin, award-winning author of *Dead Reckoning: Confronting the Crisis in Pacific Fisheries* and *The Last Great Sea,* a critically acclaimed natural history of the North Pacific,

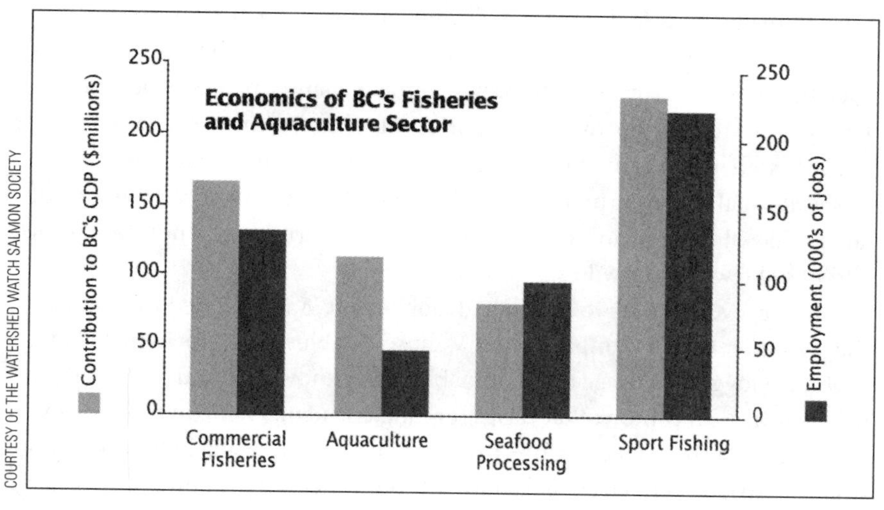

COURTESY OF THE WATERSHED WATCH SALMON SOCIETY

observed that similar values were not only reflected in an array of public opinion polls, but were also put into practice on a daily basis by volunteers working in a variety of salmon conservation initiatives.

In fact, Glavin pointed out, a public opinion survey conducted in the spring of 2000 by the DFO's Habitat Conservation and Stewardship Program found that "support for conserving salmon habitat, even at the expense of economic development, ranged from 61 percent in north coastal BC, where the economic aspects of salmon fishing are most important, to 72 percent in the Lower Mainland, where the economic value of salmon is least significant." Support for the idea of paying higher taxes if necessary to protect the fish and wildlife habitat ranged from 58 percent of respondents in the southern Interior to 69 percent in BC's south coast areas.

"Most British Columbians rank the commercial value of salmon behind a range of other values," Glavin said. "These included the contributions salmon make to ecological health, to the 'beauty of the region' and to tourism, recreation and the enhancement of 'community involvement'."

These values coincide with those expressed in national opinion surveys. A Pollara poll conducted early in 2001 found that 94 percent of Canadians wanted laws to protect endangered species in Canada, a level of support Glavin identified as an anomaly for such surveys, which rarely find public support exceeding 90 percent for any issue. The overwhelming majority of respondents to this survey—86 percent—agreed that endangered species protection should take priority over economic development.

So it was perhaps understandable that Walling and the industry would want to make the case that salmon farming as it was practised in BC was not responsible for either the marked declines in some wild Pacific salmon stocks over the previous decade or the catastrophic collapse of pink stocks in the Broughton Archipelago. In fact, supporters of salmon farmers suggested these declines were likely caused by a combination of climate change, overfishing, freshwater-habitat destruction and the genetic and ecological impacts of large-scale salmon enhancement projects. Salmon farming was not the culprit and posed few risks to wild salmon.

In any event, it was to facilitate dialogue that Walling asked me to make the icy winter trip to Campbell River, a muscular, blue-collar town that prides itself on a logging heritage but is arguably better known around the world for the gigantic wild chinook that earn sport anglers membership in the exclusive Tyee Club—if they can catch one from a small rowboat using light tackle and trolling a strictly defined artificial lure without benefit of the downriggers, fish finders and global positioning systems that so debase most of the contemporary "sports" fishery.

Walling had mustered some of the leading advocates of salmon farming on the coast to offer the industry's perspective on some of the key issues now driving the public debate. Present were David Groves, a Purdue-educated biochemist and animal nutrition specialist who is one of the pioneers of finfish aquaculture on the West Coast and whom I'd interviewed earlier during my foray to Phillips Arm, and Don Millerd of Batchelor Bay Management Ltd., whose processing plant at Brown's Bay I'd visited earlier, and which he said provided 150 steady jobs in a community hit hard by downturns in both forest and commercial fishing sectors.

Linda Sams was the senior biologist with Marine Harvest, a subsidiary of Nutreco, one of the world's largest aquaculture companies. She had helped the Kitasoo set up their contentious fish farm operations at Klemtu, an experience she described as an education in First Nation perspectives, although the Native fish farm remains a source of irritation to Heiltsuk and Nuxalk neighbours at Bella Bella and Bella Coola.

Odd Grydeland was a former faller from the woods who had reinvented his career in aquaculture and was the strategic development manager for the West Coast operations of Heritage Salmon Ltd., which operated fish farms at the epicentre of the Broughton Archipelago controversy.

They were to speak on behalf of an industry that contributes $600 million a year to BC's economy, providing more than 1,700 jobs in coastal communities beyond the urban heartlands of Greater Vancouver and southern Vancouver Island. Farmed salmon today accounts for about fifteen percent of BC's total agricultural output and ships close to 75,000 tonnes a year, sufficient to rank the province as the world's fourth-largest producer. Even at that, the fledgling BC industry is swimming with cutthroat competition. In 2000, world production of farmed salmon topped one million metric tonnes for the first time, with more than half the production coming from two countries, Norway and Chile. Third place went to the United Kingdom, with more than double Canada's production.

The gathered leaders of BC's fish farm industry were careful to point out their own concerns and attachments to the old ways. "I spent my lifetime in the fish business," said Millerd, whose grandfather owned a cannery in West Vancouver where the DFO's research labs are now located. "My first ambition was to get on the boats."

Millerd said he began shipping out as crew on north-bound fish boats at the age of fifteen and "fell in love with the coast and the fish business." He was on a business trip marketing chum salmon in Europe, he said, when he encountered a competitor marketing a fish he'd never seen before. "Every filet was perfect. He told me it was farmed salmon. That product was available year round." And that ability to put fresh product into the market regardless of the seasonal cycles governing the wild fishery was the most convincing argument that a new way of supplying fish had arrived.

Wild fish, Millerd pointed out, are governed by cycles of glut and scarcity that cause the price to fluctuate wildly. When the commercial fleet has a good catch, fish flood the market, driving prices down, often to less than subsistence value for the harvester. The wild fish can be frozen and brought to market in the off-season, but the market for frozen fish is collapsing in the face of the availability of fresh farmed fish.

"Our industry, like any new activity, has made mistakes," Millerd said. "I believe that it's improving. I think sea lice is a good example of that. It's a wake-up call. There's an environmental impact, but we truly believe it is a modest impact. At the end of the day, I'm proud of what I do."

According to Groves, the sea lice problem in the Broughton Archipelago is more complex than the European models of the parasite's impact that some fish farm critics cite. He said that pink runs into the Broughton Archipelago had increased in every even year since 1974, and that for the decade prior to 1998, a period during which fish farms were present, there were six successive cycles in which spawning escapements averaged 1.5 million fish, resulting in the record return of 3.5 million pinks in 2000.

Groves said the success of the pinks might, in fact, be the underlying cause of the crash. While the feed supply for wild smolts was fixed by the carrying capacity of the ecosystem, the biomass of hatching salmon could have overwhelmed its food source. As an animal nutritionist, Groves said, it's not difficult to calculate the amount of feed required in a normal cycle and to compare the amount needed in a record year to the amount of feed available. From one year to the next, for example, the number of migrating smolts doubled from 240 million to 500 million, he said. Such numbers simply stripped feed out of the system.

"What happens when you double the stocking level in an area where there's already only twenty percent of the food necessary?" Groves asked. "There had to be a lot of competition [for feed]. You probably had 85 percent of all the smolts in Tribune Channel and they just hoovered up the food supply. If there are sea lice there, these little fish were increasingly starving fish and were more vulnerable to the lice."

Groves suggested that even if sea lice clearly need to be managed on fish farms, wild salmon stocks also need to be managed to prevent the kind of system overloads that occurred in the Broughton Archipelago. "We can never allow such a horrendous number of pink salmon into the Broughton again," he said. "That's wild salmon management. To raise salmon, we are managing not just the fish in our nets. We have to manage the surrounding environment. That was the thing that really tipped it—there were too many fish in the system."

On the other hand, he observed, the smaller number of pink smolts migrating the following spring would have access to a superlative feed supply. And speaking of feed, I asked, what about the argument raised by some academics and environmentalists that fish farming is not sustainable because it requires increased harvesting of other stocks to provide the pellets used in aquaculture, a process that's been described as fishing down the food chain?

Millerd countered that the statistical evidence doesn't support the case. "I could see that criticism if the reduction fishery were increasing worldwide—but it's not," he said.

In fact, Groves argues, aquaculture worldwide uses only 40 to 50 percent of fish meal production. The rest goes to feed swine and poultry, which are much less efficient at converting protein to an edible form used by humans. What's more, he said, the harvest for fish meal is about the same as the unused bycatch in global marine harvests, the bulk of which is simply tossed back over the side.

And so, point by point, the salmon farmers offered answers for their critics:

- Antibiotic use in farmed salmon is minuscule compared to its use in other livestock like cattle, swine and poultry.
- Sewage from fish farms is minuscule compared to the billions of litres of human sewage discharged into the ocean from Vancouver, Victoria and other urban centres each day.
- Opposition from aboriginal leaders doesn't take into account the positive benefits of steady employment in those First Nations communities that have embraced fish farms.
- Escapes of Atlantic salmon have been reduced to a marginal level and those that do escape are likely too small to compete with wild Pacific salmon and survive.
- Diseases like the IHN virus, to which Atlantic salmon are extremely susceptible, remain a problem, but efforts to find a vaccine continue, and in the meantime a huge fish health database will track and report every mortality.

It all sounded convincing, but back in my hotel room at the Tyee Plaza, as I organized the stack of papers, reports, press releases and three-ring binder bestowed on me by Walling, my folder of background notes spilled some recent newspaper clippings that didn't foster reassurance.

One pointed out that BC got a failing grade for inadequate research of fish health issues. Another reported that under the present provincial government, BC's spending on wildlife protection had collapsed to what it was twenty years earlier, while spending on enforcement had been halved. A third quoted scientists saying a fundamental change in government attitude must occur if there is to be a solution to endemic disease problems on salmon farms. And the latest, barely a week old, quoted Kwakwaka'wakw chief Bill Cranmer's complaint about "sickening" conditions at a disease-infested salmon farm in the Broughton where "fish guts, fish oil and diseased fish were being discharged into the water"—all within the existing hygiene guidelines. Even the province's sustainable resources minister, as staunch a defender of salmon farming as could be found, told the weekly *North Islander* newspaper, that what he saw on a tour of the Broughton site was unacceptable by present standards.

Those clippings served up a sharp reminder that, as in every controversy, a sincere belief that your facts are the right facts doesn't necessarily negate your critics' arguments—or in this case allay their fears for the future of BC's wild salmon.

The Sport Anglers

A sullen swell the colour of old pewter catches the first light spilling across foam-edged shoals at Race Rocks. The freighter *Barnard Castle* was opened like a tin can on those reefs back in 1886, although the occasional elephant seal and the even rarer fur seal seem to be able to hump their way over the sharp edges with impunity.

This is the real skookumchuck, far too much ocean for my little aluminum runabout. The chart shows twelve tide rips off Cape Calver at the southernmost tip of Vancouver Island, some of them surpassed in violence only by treacherous Seymour Narrows near Campbell River.

I'm out here with fishing guide Gord Gavin of Sea Ghost Charters for two reasons. First, to get a feel for the often overlooked recreational end of the debate now raging around the question of fish farms and their potential impact on wild salmon. Second, to give my daughter Heledd her first attempt at every British Columbian's birthright—catching a big winter spring with a herring strip on a bitter morning of mist and rain squalls.

There are few things more exhilarating than listening to my old single-action Peetz reel sing as a tyee makes his straight, true run. The reel is made of wood and bronze. It was my father-in-law's more than half a century ago, and it will be Heledd's one day, when she's ready to take care of it. Just taking it out of the old purple velvet Crown Royal whisky bag evokes a flood of memories, a ghostly latticework of stories that reach back to the very origins of this place.

Recreational angling for salmon is likely this province's oldest and most broadly participatory sport. It reaches back to the early Victorian era, an egalitarian pursuit that cuts across the class boundaries which define such fishing elsewhere as a preserve of the privileged—although wealthy anglers have also beaten a path here ever since a 19th-century Royal Navy officer wrote "A Seventy Pound Salmon with Rod and Line" for the London sporting magazine *The Field.*

Now, as fish farms proliferate along this coast—a private interest colonizing the public commons, trailed by a slick of untreated sewage, toxic algae blooms, disease outbreaks and truck-sized totes of dead fish (morts, they call them, in one of those creepy industrial euphemisms), escapes of exotic species by the thousands, the fish farmers' slaughter of seals and sea lions and allegations that sea lice concentrating around netpens kill migrating salmon smolts—some wonder whether the people's ancient sport can endure.

Gavin runs about 80 charters a year, and on this cold, crisp morning, as we leave Race Rocks behind and turn into Whirl Bay, he tells me he's been catching winter springs of anywhere from three to fourteen kilograms. "It's been really good this winter and I suspect it will be good for the rest of the summer," he says. "We've had a strong run of springs, both winter and summer. I think the reduction of commercial netting up the west coast has had a positive impact on returns. Six years ago, when they closed the coho fishery, there was a big fuss. But [then federal fisheries minister David] Anderson was right—you have to look at the long term. That fishery has come back very strong."

I ask him what he thinks of the commotion over pink salmon runs in the Broughton Archipelago and the suggestion that sea lice from fish farms devoured wild smolts on their outward migration.

Gavin, the son of a Prince Edward Island commercial fisherman, mulls the question while he checks the gear, showing Heledd how to rig an anchovy on one line, a herring strip on another and the fluorescent squid imitation called a hootchie on a third.

"I think they kind of rushed into this fish farming thing without fully knowing what they were doing," he says. "I've been following it for quite a few years now. There was the moratorium on further expansion and I thought they'd have been doing studies to determine what could be done to avoid problems in other areas. It seems that hasn't happened—it's a different government with a different approach, I guess."

Tom Bird, executive director of the Sport Fishing Institute of BC, echoes some of those views when I track him down in Richmond. Bird, who describes himself as the last of the gumboot biologists, served thirty years with the DFO, twenty of them in habitat protection and ten as director of recreational fishing. The last five years he's been with the institute, which represents marina and fishing lodge operators, boat dealers, equipment manufacturers and retailers of tackle and bait, all those elements of the vast economic infrastructure that depends upon healthy wild salmon stocks and sport anglers determined to catch them.

Generally, Bird says, the institute prefers to work behind the scenes rather than wage policy battles in the media. He's also quick to say that fish farming as a concept is sound and that in a world where population is growing and global fish stocks are depleted, it's probably inevitable. "But the risk to wild stocks—well, then it has to be bulletproof, absolutely bulletproof. We cannot jeopardize the wild stocks. We simply cannot do that."

It's pretty easy to understand his intensity. BC's recreational fishery is a major driver for the province's $1.5 billion tourist industry. Indeed, Bird points out, the institute's members promote sport angling in the context of BC as a tourist destination, and "the economic returns are very large for a minimal impact on the resource."

According to Statistics Canada, more than half a million anglers fished here in 2000. They spent a total of $1.2 billion on their pastime, more than double the total value of BC's fish farm production in 2002. Anglers spent $163.6 million on fishing tackle and boating gear alone, almost five times the $34 million in wages paid by all BC's aquaculture industries.

These are astonishing numbers. And the multiplier effect must be vast in small coastal communities when compared to the jobs generated by the salmon farming industry, which in other countries has actually seen employment falling as automation accelerates. BC's aquaculture sector employed only 1,900 people in 2001, says BC Stats, compared to 8,900 jobs in the recreational fishing sector. In Scotland fewer than 2,000 people were employed by fish farms in 2000, although production increased by almost 10,000 tonnes. In Norway, while production was almost doubling, fish farms shed more than twenty percent of their workers, falling to 2,346 jobs in Atlantic salmon and rainbow trout operations by 2001.

So it's surprising to some that the provincial government has rushed into private fish farm expansion without having adequate science to provide those "bulletproof" assurances demanded by sport anglers, who represent a huge public constituency and yet feel neglected.

"I have no answer," Bird says. "It doesn't equate."

He does confirm that concern about fish farms and potential risks for wild fish is growing. Partly, perhaps, this is because fly-fishing for pink salmon in tidal estuaries is emerging as one of the dramatic growth sectors in the sport fishery. "The pink salmon is becoming increasingly important to the sport angling sector," Bird says. "Beach angling for pinks and coho is just exploding. On the beach by my place there used to be two guys out there. Now I have to fight for a place."

The emerging sea lice problem for pink smolts running the gauntlet of fish farms in the Broughton Archipelago lies at the heart of current concerns of sport anglers. "We all know that sea lice are everywhere," Bird says. "We know that it's a natural occurrence for sea lice infestations in the wild. We also know that there is some kind of relation. Our members are vitally concerned. We have a pressing concern that the investigation be aggressively pursued."

Worries about the failure of the precautionary principle that inadequate science represents are expressed in a draft resolution by the institute calling for an end to netpen salmon aquaculture "unless and until the federal and provincial governments can demonstrate that this form of salmon farming can be operated in a manner that does not pose a threat to naturally reproducing salmon stocks."

The late Lee Straight, who spent 33 legendary years as the *Vancouver Sun*'s outdoors columnist, was less diplomatic when I talked to him not long before he died. At 87, having witnessed the cycles of abundance and decline for himself, Straight described Alexandra Morton, who is routinely vilified by critics, as "a heroine."

"I'm against fish farms per se," he said. "I think they haven't been well enough researched. It's just like all these experimental things. We've rushed in without fully realizing all the consequences. I want to see them all shut down. Right now, too. I'm sure the sport anglers feel the same way in Norway, Scotland and Ireland."

Van Egan, a biologist, teacher and writer who has lived beside the Campbell River for 48 of his more than 76 years, and fished and guided as a contemporary of internationally renowned naturalist Roderick Haig-Brown, is equally blunt.

"Shut 'em down," says the author of *Tyee: The Story of the Tyee Club of B.C.* "That's the best thing that could happen to our wild salmon right now. I would phase them out. Since they've refused to go into land-based containment systems, I think we'd be better off if they just went to Chile. They have to be environmentally neutral and I don't think that's possible by having net-pens out in the open water being used by wild fish.

"One of the things that most bothers me is the fact that our local papers are being inundated with letters from people who supposedly work for the fish farms," Egan says. "You get letters pronouncing the economic benefits and the jobs that would be lost by making fish farms environmentally neutral. What about the jobs that might be lost in the sport fishery or the commercial fishery by letting them use the environment the way they do? They are always talking about good hard science, about the lack of science that shows a direct impact. There's never much talk from them about the absences of the hard science proving that they don't have any impact on wild fish. To me there's plenty of science and some of it doesn't even need science—a high school student could figure out the answers to some of this stuff."

Peter Broomhall, a well-known BC angler and long-time member of the lofty Totem Fly Fishers, BC's senior fly-fishing club, says he's not against fish farming any more than he's against logging. But he believes it shouldn't be done at all if it can't be done properly—and he says it's not being done properly.

"We are not using full cost accounting when we look at these fish farms," he says. "We loosen our environmental laws to accommodate the needs of industry. This seems an insane way to go about it. If these guys can't do it without stressing the environment, they shouldn't be doing it. They've had a free ride. That's got to stop. If I could, I'd reinstitute the moratorium. I'd say, 'It's wild salmon first and anything that endangers them must come second.' What astonishes me is that we've found it so easy to cavalierly hand over the resource to the plunderers."

Out on Whirl Bay, all we've got for our trouble so far is a few mangled anchovies. Gavin is taking his lines across an eddy and down the edge of a hidden reef while he explains that he thinks guides like himself should be licensed when Heledd's rod bucks and the reel screams.

Then it's up to her, and she does just fine, rod up, line tight to keep the barbless hook set, playing her first wild winter spring to the boat. It's nothing like the tyee that left the European sporting establishment agog more than a century ago, but it's a nice dinner-sized fish and she'll keep it.

A few minutes later we catch one more of the province's defining symbols—and that's plenty for us. We pack up the gear and run for home, one more BC girl having claimed her birthright, one more proud yet slightly melancholy dad wondering if she'll be able to savour the same moment with her own child somewhere in the uncertain century ahead.

Conclusions

The sandstone ledges that step down into the deep green current of the Nanaimo River look smaller, narrower and less awesome than I remember, although the bank above York's Hole seems steeper and more dangerous now than it did when I clambered down it as a six-year-old.

That long, lazy, sun-filled summer I'd swim across to the big gravel bar, wade into the fast shallows and stand there mesmerized while pink salmon preparing to spawn by the thousands milled around my bare, skinny legs. In this river just south of Nanaimo I caught my first salmon, bigger in my recollections than a proud father's photographs indicate, undoubtedly because I was so much smaller then. Today the river is closed to anglers.

Things change in half a century, not least one's perceptions. Across the river, what used to be the Polkinghorne farm has dwindled to Lilliputian proportions. Among the biggest changes on this coast have been the rise of the salmon farming industry, the decline of the commercial fishery and, although sometimes it seems only the oldest of old-timers have noticed, the decline of the sport fishery.

Before 1985, British Columbia produced fifteen percent of the world's salmon, virtually all of it from wild stocks. By 2000, commercial landings of wild fish had declined by 80 percent while output from salmon farms was up by a similar percentage, more than 80 percent of that from an exotic introduced species, the Atlantic salmon. The gap between wild and farmed production seems certain to grow with the lifting of the province's moratorium in September 2002, accompanied by what seems a relentless determination to see big corporate salmon farming expand even as government's commitment to wild fish diminishes, assurances to the contrary belied by round after round of environment ministry budget cuts, staff reductions and impoverished field operations.

And yet some things don't change.

Hidden in that translucent sweep of water, beneath the sun sparkles skittering across its surface riffles, the timeless comings and goings of BC's wild salmon are still taking place. For while it may surprise some, the Nanaimo River is home to every native anadromous salmonid—even a tattered remnant of sockeye—and there is not a month in which at least one species is not migrating. Outbound, the ruined pink and still-abundant chum fingerlings ride the spring rains toward the sea. Inbound, a unique race of slab-sided chinooks heads up the river, where it will spend the summer in the Nanaimo Lakes before turning downstream to spawn in September.

All along the coast from San Francisco to Old Crow, deep in the interior of Yukon, hundreds of millions of these amazing creatures are in motion, weaving their simple patterns of life and death into the vast, complicated ecology of what essayists Elizabeth Woody, Jim Lichatowich, Richard Manning, Freeman House and Seth Zukerman called the Salmon Nation in a remarkable book of the same name published by Ecotrust in 1999. They defined the nation as that region on the northwestern slope of North America where the cycle of natural abundance that wild fish represent is embedded in the cultural identity of whole peoples—and species, too, from grizzly bears to bald eagles.

So why do I revisit this fragment of the Salmon Nation retrieved from my distant childhood in order to contemplate the present controversy surrounding salmon farms? Because it's important to take the long view rather than accept the four-year horizon of politicians and the quarter-to-quarter perspective of business managers.

The Nanaimo River might serve as a metaphor both for our collective abuse of the astonishing richness we continue to fritter away in the service of politics, greed and short-term thinking and for the slim hope that some long-term thinking might bear fruit.

Today salmon farms pose a profound new threat to the survival of wild salmon, in no small part because of the certainty among their supporters that there's little risk. But it's important to remember that salmon farmers are only the most recent example of our willingness to abuse what some see as a divine gift from the Great Spirit. Miners, loggers, commercial fishermen, industrialists, real estate developers, urban planners, dam builders, road engineers, careless householders, politicians bereft of vision and scientists without courage, each prepared to tolerate some specific demand on habitat, have left their ugly incremental mark on the Salmon Nation.

My journey through the recent storm of charges and denials about the risk from fish farms began in the Broughton Archipelago, far to the north. There I found wilderness rivers that seemed as silent after the failure of the previous summer's pink salmon runs—a collapse that some scientists said was beyond the magnitude of anything they'd seen—as the long-ravaged Nanaimo seems today. Some research pointed directly at fish farms, well-known from the Norwegian, Scottish and Irish experiences to be breeding grounds for sea lice. Sampling by volunteers in the face of apparent indifference from DFO's scientists and bureaucrats responsible for husbanding wild stocks found evidence strongly suggesting the parasite had devoured migrating pink smolts alive as they swam through infested waters.

This was greeted by a chorus of denial from the fish farmers, from federal authorities committed to promoting the industry and from politicians in a provincial government characterized by a dogged ideological dismantling of an environmental stewardship apparatus assembled over the better part of a century—a process accompanied by cynical assurances to the contrary even as the province eliminated funding for habitat improvement.

Here on the banks of the Nanaimo River, at a place where I remember the shallows seething with spawning fish that summer of 1953—the same summer I watched the Snuneymuxw women with their long-poled gaffs and woven baskets taking their share of the bounty—seems as good a place as any to consider the self-interested certainty of industry, the credulous greed of governments and what these attitudes might imply for the natural world.

Today this is a river on life support, its once-teeming wild runs kept from dwindling into extinction only by the Nanaimo River Salmonid Enhancement Project.

"The Native elders tell me the Nanaimo used to be black with pinks," Patti McKay told me when I stopped by the enhancement project office. "They say they had way more pinks in this river than DFO's own figures report."

Mine tailings wiped out what was once a big run of pink salmon into Napoleon Creek, a tributary of the Nanaimo, says Snuneymuxw fisheries worker Henry Bob, who has spent 25 years on the river. The last gasp of that run, spawning in the channel below Polkinghorne farm, was what I witnessed in 1953. By 1963 the count was down to one fish. In 1973 the count was zero. The numbers crept back up to 50 fish in 1994, but by 1997 it was down to three fish, and when I clicked through the fancy templates of the computerized database set up by the federal and provincial fisheries authorities to determine the most recent status of pinks in the Nanaimo system, I got a "no data" message.

"If a species is missing from the river, there's something really wrong with that ecosystem," McKay says. "Pinks are really important. Even if only as a feeder fish [for other species] they are just so important in the estuary." Which is why, she says, the project she co-manages will attempt to begin rebuilding the shattered pink run in the Nanaimo River. "We've been trying for five years and we've finally managed to obtain [eggs] for 2004," says McKay. "It's pretty exciting."

Other salmon species suffered as well. A dam denied the upper river to chinook. Water was sucked out of the system for pulp mills and urban development. Logging in the headwaters, which is still going on, changed the runoff patterns. Coho, already suffering because of poor ocean conditions, collapsed. So did steelhead.

"Our biggest concern now is logging. Right now the sediment is the worst I've seen in the last ten years," Bob told me as I sipped his stand-up-your-spoon coffee in the enhancement project office. "We get these big freshets now and they scour out the river. The little fish just get blown out and the freshets take away the spawning gravel. There's very little habitat left for the chinooks."

Still, he says, it's because of the project that chinook returns were stabilized at about 2,800 per year—although that's less than half the normal peak—and coho were able to hang on.

Jack Whitlam, 83, worries that maintaining returns won't compensate for ongoing damage to crucial salmon-rearing habitat in the river's estuary, which is used as a booming ground. He's quick to stress that he's not against logging, just the shoddy way it continues to be done. "Now they are using log bundles instead of flat booms," Whitlam complains. "It's more efficient for the companies, but you get up to six times the debris, and these bundles are so heavy that they tear up the bottom and destroy habitat for crustaceans, insects and the eelgrass beds that provide feed for salmon and trout."

Ted Barsby was born in Nanaimo 83 years ago, and except for a wartime job flying Lancaster bombers he has lived all his life here, tramping the watersheds of the Nanaimo River with rod and rifle. He was the founding president of the BC Wildlife Federation. "Maybe I'm just too bloody old," he says ruefully. "I knew all that country when it was virgin. It's a totally different river than it was. Back in those days I could go up to the Bore Hole and I could catch chinook up to 30 pounds. Deadwood Creek was just full of steelhead.

"When I came back from the war in 1945 they were logging up there," Barsby says. "The steelhead disappeared around then. They are still logging heavily back in there. These losses are never taken into account when the government calculates the costs and the benefits. You always get a false figure."

Barsby says the fish farms proliferating in BC waters are just the latest insult to wild salmon. "What do I think about fish farms? Not much," he snorts. "They went about it the wrong way from the start. Environmentally they are just not sound. I'm opposed to them. Unfortunately, anything that will make a buck these days, that's what the politicians go for. It's the fast-buck mentality. Politicians, they are the ones who are messing everything up."

The skeptical perspective is remarkably similar to that expressed by Otto Langer, a former DFO scientist who now works as director of marine conservation for the David Suzuki Foundation. In a peer-reviewed paper published in 2003 in *BioLine,* the journal of the Association of Professional Biologists of British Columbia, this veteran fisheries expert doesn't mince words. Langer says open netpens used to rear salmon in close confinement are the root cause of many environmental problems, ranging from sustainability to disease and parasite amplification, which pose significant threats to wild salmon stocks already stressed by other factors.

Science notwithstanding, simple common sense warns that risks are enhanced because the crowded conditions in netpens encourage disease and parasite transmission. In addition, Atlantic salmon have shown less natural resistance to the diseases of wild Pacific salmon, which are able to disperse over wide areas, making them less vulnerable to both the spread of contagion and infestation by parasites. This is a key factor in the inability of the salmon farming industry to rid itself of persistent outbreaks of disease, the most virulent of which is infectious hematopoietic necrosis (IHN).

It has been difficult to pry historic data about disease mortalities out of the province, but I do learn that between August 2001 and May 2002, IHN outbreaks killed millions of salmon at nineteen fish farm sites, most located where they created major pools of infection right along the migration routes for wild salmon. And later in 2002 the Broughton pink runs collapsed, just as Alexandra Morton predicted after she found migrating smolts covered with sea lice in proximity to salmon farms.

Other voices were raising warnings about the lack of scientific knowledge. Back in June 2001, more than a year before BC's provincial government decided to lift its moratorium on salmon farm expansion, the Senate's standing committee on fisheries released a report, *Aquaculture in Canada's Atlantic and Pacific Regions,* based on hearings across Canada that had warned the environmental science on potential impacts from netpen aquaculture was inadequate. The Senate report said that not enough was known about possible ecological and genetic effects of escaped farmed salmon on wild species, the interaction of fish farms with aquatic mammals and other species, the incidence and transfer of disease in farmed and wild stocks, and the environmental risks associated with the wastes discharged by farms.

"Without sound scientific knowledge, it is difficult to see how regulatory agencies can set meaningful standards, guidelines and objectives," the Senate committee observed. "Without such information, suspicion and distrust of the industry will continue." It recommended that salmon farms simply be prohibited near migratory routes as well as near any rivers and streams that support wild salmon stocks.

In August 2002 a Raincoast Conservation Society team of seven biologists, ecologists and geneticists examined databases on returning salmon over the previous three years and found no information on stocks for 70 percent of the 2,500 salmon runs in the area surveyed. In its report "Ghost Runs: The Future of Wild Salmon on the North and Central Coasts of British Columbia," the team urged that no further expansion of fish farms take place until more scientific knowledge could be acquired.

In spite of this dearth of reliable information, the BC government lifted the moratorium on expansion, claiming it had all the science it needed to support its position. Six months later, the editor of a report on the IHN virus that was ravaging Atlantic salmon in netpens on the West Coast warned that BC actually lagged far behind other jurisdictions in scientific research on fish health issues. In retrospect, considering the ongoing health crisis for farmed Atlantic salmon in BC waters and the clear potential for adverse and possibly irreversible interactions between farmed and wild salmon, my travels had helped clarify one opinion—the provincial government's decision to lift the moratorium was ill-judged, irresponsible and apparently motivated by ideology rather than intelligence.

The argument from critics that the moratorium should be restored until there's convincing scientific consensus about how to proceed safely seems reasonable. Fish farms that clearly pose a risk to migrating wild stocks, either from disease or parasites, should be closed or moved. And politicians need to hear the message that if they cannot demonstrate their commitment to the principles of stewardship, they can expect to be dismissed by the voters.

Standing beside the ravaged Nanaimo River, contemplating the recurring pattern of abuse and denial that characterizes our failed duty to preserve the bounty with which we've been blessed, I'm left only with questions where I came looking for answers: Can we save our wild salmon, that symbol of everything that is free and untamed? Is it too much to expect the kind of moral leadership that can insist industry prove beyond the shadow of a doubt that its enterprise is safe and sustainable? Or will our wild salmon go the way of the buffalo and the passenger pigeon, one more sacrifice at the altar of greed and recklessness?

Politicians mesmerized by the prospect of new revenues, scientists whose careers depend on federal and provincial bureaucracies that are ideologically committed to aquaculture, and industry executives driven by their balance sheets repeatedly promise that farmed fish are no risk to wild salmon and that the two populations can coexist. Yet even as they offer these assurances, some of the wild stocks they are responsible for protecting continue to spiral toward oblivion. In 2003, sockeye from Sakinaw Lake near Sechelt and Cultus Lake in the Lower Mainland were declared endangered species, and it's estimated that 232 distinct salmon populations have already become extinct from BC to California.

Now is the time for more of us to assert ourselves on behalf of our wilderness, parks and wildlife. Salmon farmers and other industries need to hear the message loud and clear: Operate safely, sustainably and disease free and prove there is no risk to wild salmon stocks or take the business somewhere else.

After just 50 years, the beautiful river of my childhood is a shadow of its former abundance. But perhaps, just perhaps, if we hold our business and political leaders accountable, we can still ensure that both the natural world and British Columbia's future children inherit a richer legacy.

Sea-Silver

A Brief History of British Columbia's Salmon Farming Industry

Betty C. Keller and Rosella M. Leslie

On the BC coast today, salmon farming begins at a company hatchery located where fresh water can be diverted from a nearby stream or where pure underground water supplies can be tapped. During the fall, eggs are collected from mature female salmon or "broodstock" that have been selected for their growth potential, feed conversion and maturation rates, and resistance to disease. This "egg intake" is screened for disease, fertilized with milt taken from selected males, treated with a surface disinfectant, then placed in incubation trays. Because salmon are a cold-water species, the optimum temperature range for the water flowing through the incubation units is 5°C to 16°C. Lower temperatures slow growth; higher temperatures reduce disease resistance.

After approximately 30 days, the eggs reach the "eyed" stage, which indicates that they are alive, and workers cull the dead and unfertilized eggs. Alevins or yolk-sac fry hatch out 40 to 50 days after fertilization; they are kept in the incubation units for a further 20 to 30 days until the fry have completely absorbed their yolk sacs and weigh between 0.25 and 0.5 grams. At this point they are transferred to freshwater rearing tanks, where they are fed intensively until they become smolts—the stage at which they are physiologically capable of living in seawater. The smolts are moved to saltwater netpens, using a combination of truck, barge and "well-boat" transport systems, sometime between March and August, depending on the species.

The saltwater farms where the smolts will be reared to market size are generally sited where an island or headland will give some shelter from the rougher weather, but where water circulation is vigorous. Each farm consists of a system of interconnected netpens anchored to the sea bottom. The surface measurement of the average pen used on the BC coast is fifteen metres square, and it can be up to ten metres deep. The average stocking density for Pacific salmon species within the pens is eight kilograms of fish for every cubic metre of space, while Atlantics, which by 1995 were being reared at 70 percent of BC's farms, are stocked at approximately ten kilograms per cubic metre. Ideally, farm companies have several sites set aside for their smolt intake in order to leave each site fallow for a year between crops, thereby lowering the risk of disease being transferred from the previous crop.

The fish are then fed in a carefully regulated manner until they reach marketable size, usually in the neighbourhood of eighteen months. If all goes well, it is a fairly straightforward and uneventful process. But salmon farming on the BC coast has been anything but uneventful over the years as it has been continuously plagued by setbacks and public controversy.

Salmon farming differs radically from other agricultural operations such as the farming of sheep, chickens or cattle in one major respect: in agriculture the term "farm" refers to the piece of land used to raise crops or animals, while in aquaculture "farm" means only the system of pens in which the fish are reared and their attached, barge-mounted operations centre. The site of the farm is publicly owned water that has been leased from the government for aquacultural use. However, the water that surrounds the farm has the potential to carry the effects of the farm operation far beyond the boundaries of the licensed area. This fact is at the heart of the friction that has developed between the industry and the general public in British Columbia.

Many of the problems arose because the industry was initially regulated by four federal departments and five provincial ministries (all of which changed names and responsibilities during these years), which led to as much jostling for political turf as careful science and regulation. From the beginning, the Department of Fisheries and Oceans (DFO) was the federal government's main arm for dealing with aquaculture, although until the 1990s the department's stated mandate had been management of the wild fishery. The department's main adversary at the provincial level—where the pro-business Social Credit was in power—was the Ministry of Agriculture, Fisheries and Food (MAFF).[1] It in turn vied with ministries such as Environment (MoE) and Lands, Parks and Housing (MLPH) for control over the development of aquaculture. Unfortunately, the policies guiding all of these ministries and departments were developed by doctoring policies and acts intended to regulate wild fisheries and/or agriculture, not to regulate an industry that is essentially agriculture in an aquatic environment.

BC's salmon farming industry was left to develop for fourteen tumultuous years in this vacuum. Then in 1986, after massive fish losses from algal blooms and disease, and as complaints from the public increased, the provincial government announced a moratorium on new site licences and set up a commission to reorder the industry. After a month-long inquiry, a simplification of licensing procedures and the addition of a few new regulations, it was business as usual.

Conditions within the industry changed dramatically during the following eight years. In 1989 there were 185 small salmon farms in BC waters operated by more than 100 companies; by 1993, as a result of storms, disease, algal blooms and rock-bottom salmon prices, those numbers had shrunk to 80 farms operated by seventeen companies, but they had become large farms and international companies. With fewer employers had come fewer jobs. The industry's focus, which had been in Sunshine Coast waters in the 1980s, moved north and west to avoid the warmer waters where algal blooms flourished. Pacific salmon were mostly replaced by Atlantic salmon, and new, more deadly environmental concerns arose.

In the mid-1990s, both federal and provincial governments, responding belatedly to public pressure, were forced to act to reorder the industry yet again. In April 1995 the federal government announced a new aquaculture development strategy: "Aquaculture development must not be unduly constrained or burdened by government policy or the regulatory framework. At the same time, however, aquaculture development must be consistent with government responsibilities in such areas as habitat and biodiversity." The federal document went on to promise new, commercially oriented research programs, access to new technology, new training programs for workers in the industry, promotional support and "equitable access" to aquatic resources.[2] The industry applauded the federal government's move.

In May 1995 the provincial government, now led by the more environmentally conscious NDP, released its *Action Plan for Salmon Aquaculture in BC,* which was intended to address five broad areas of concern: the farming of Atlantic salmon, waste management, disease transfer, standards of farm operation and the formulation of standardized policies. At the same time, it announced that an "unofficial moratorium" on new fish farm licences, which had been in effect for almost eighteen months, would remain unofficially in effect for at least another year while studies were carried out to determine how to prevent or reduce the adverse effects of salmon farming. In particular, BC's Environmental Assessment Office would conduct an environmental review of

the salmon farming area surrounding the Broughton Archipelago, where farms belonging to BC Packers and Stolt Sea Farm Ltd. were licensed to operate.

The industry's reaction to the NDP government's action plan was frustration and anger. Mike Mulholland, president of the BC Salmon Farmers Association, told the *Globe and Mail*'s Robert Williamson that the industry's largest companies would be "steering millions of dollars in job-creating investment outside Canada to rival nations, including Chile and the United States."[3]

In fact, BC Packers, Stolt Sea Farm Ltd., Moore-Clark Ltd. and International Aqua Foods—the four companies responsible for 60 percent of BC's farmed fish production at that time—already owned large farms in both the US and Chile. Investment in the US was an advantageous move for BC companies since salmon raised in the US did not have to pass border inspections if they were also sold there. Raising salmon in Chile had the advantage of producing fresh fish at a time of year when North American fish were not available. In addition, in the mid-1990s Chile, with its lack of regulations and restrictions on fish farming, was very investment-friendly. However, Dr. Julian Thornton of Microtech International of Victoria, which produced vaccines for the industry, cautioned that he "wouldn't want to step one foot toward their [the Chilean] method of production. They'll wake up. They'll have their problems and they will crush their industry."[4]

At the beginning of 1996, with the battle lines drawn, BC's salmon farming industry awaited the outcome of the conflict. The key, however, lay not so much in the review process, but in what had gone before in the history of the industry in this province.

The Pioneers

The first Canadian saltwater salmon farm was built in 1967 near Halifax, Nova Scotia, by Sea Pool Fisheries Ltd. It was designed to rear speckled and rainbow trout, Atlantic salmon and two species of Pacific salmon—coho and chinook. Sea Pool confidently forecast an annual harvest of 1.8 million kilograms of fish by 1971, but the farm failed to achieve even five percent of that amount. The fish died from a vitamin-deficient diet, serious overfeeding, high acid levels from fish excrement, incorrect salinity levels, inadequate freshwater supplies and endemic diseases as well as diseases imported with fresh stocks of eggs. With mortalities steadily climbing and predator birds helping themselves to the product, the company could not keep an accurate inventory or fill orders. In 1972 it was forced to declare bankruptcy.

Washington state's first salmon farm was a 1969 joint venture by the multinationals Union Carbide Ltd. and Singer Ltd. Their pioneering subsidiary, Ocean Systems, worked closely with the Manchester Marine Station, Washington State's fisheries department and the Washington Sea Grant Program, which funds marine research and development in the state's universities. Ocean Systems' pilot study was aimed at producing "pan-sized" or single-serving fish (226 to 340 grams). During the first season of operation, Ocean Systems discovered that coho eggs reared at 8°C to 10°C would smolt in less than six months; they could then be transferred to seawater pens and achieve pan-size in twelve to eighteen months. With this encouragement, the company located a major facility with Japanese-style floating netpens just across the bay from the Manchester station, and in 1972 the new farm was able to harvest over 540,000 kilograms of pan-sized coho. This success prompted Union Carbide (which by this time had bought out Singer's interest) to transfer the project from Ocean Systems' aegis to a production company called

Domsea. With Union Carbide's enormous resources to keep it afloat, the company solved its biological and engineering problems one by one.

Salmon farming on Canada's West Coast was initiated in the fall of 1971 by the forestry giant Crown Zellerbach, which installed two floating netpens about a kilometre offshore from its Ocean Falls pulp mill at the head of Cousins Inlet. One pen was devoted to chinook salmon, the other to rainbow/steelhead trout. All the fish had been obtained from a hatchery in the northwestern US. In the spring of 1972, when DFO became officially aware of the presence of these foreign fish, the culprits were ordered deported to their place of origin. No American hatchery would admit to having supplied them, however, and the Canadian paperwork had disappeared when Crown Zellerbach closed its Ocean Falls mill. It has been reliably suggested that the fish met their end on the dining tables of Ocean Falls.

Around this time, marine biologist R.J. "Roly" Brett, the senior salmon farming authority at the DFO's Pacific Biological Station (PBS) in Nanaimo, recommended that the federal government should wait five years before licensing any marine fish farms. This would allow PBS scientists to complete studies on site selection, property rights, engineering, environmental concerns, nutrition, bio-energetics, disease and predator control, processing and marketing. Brett also urged the establishment of an experimental fish farm at Nanaimo. Although the decision to build it finally came in 1973, the government did not wait to begin licensing fish farms.

The first salmon farm licence granted by the DFO went to Allan C. Meneely on June 6, 1972. Moccasin Valley Marifarms was to be based on Meneely's privately owned twelve-hectare parcel of land located close to Earl's Cove on Agamemnon Channel, at the north end of the Sechelt Peninsula. He was licensed by the province to draw water for his hatchery from the creek that flowed from North Lake through his property and was given a foreshore lease on the adjacent salt water where he would anchor his netpens.

Meneely received some advice on disease control from PBS, but he and his two sons had to rely on information from books and pamphlets and on their own engineering skills to develop their farm. As a result, although the Norwegians, Scots and Japanese had been using sophisticated netpen technology for years, it was three years before the Meneelys perfected an effective saltwater enclosure to hold Moccasin Valley's fish. Until then, the farm went from disaster to disaster, including the failure of a seven-metre-diameter fibreglass tank that burst, spewing fish fry all over the ground.

By far the most serious problem the Meneely family faced was procuring eggs for their hatchery. Since it was against federal law to import eggs except under permit, Meneely had to rely on eggs from government hatcheries. At first this seemed to pose no problem. On August 25, 1972, he was assured by the provincial Ministry of Recreation and Conservation that 150,000 coho eggs would be available through the Commercial Fisheries Branch. By mid-September he was advised that only half this quantity would be available since so much stock was required for salmon enhancement programs. When Meneely applied for a permit to import the balance, the federal Fisheries Department vetoed it on the grounds of the possible importation of diseases. The battle to obtain eggs continued year by year.

In the meantime, University of Victoria biochemist and salmon farm consultant David Groves had initiated his own salmon farming operation, Apex Bio-Resources, with partner John Stavrakov. They obtained a trout farming licence and began building a hatchery on private farmland at Westholme, just north of Duncan on Vancouver Island. By 1973–74 the hatchery was in production and Apex had been granted salmon farm licence #2 for a salmon grow-out site at Genoa Bay inside Cowichan Bay. They moored their netpens right beside a marina and began their hatchery operations with chum, sockeye and coho. While the sockeye in the PBS land pens had grown bigger and faster than wild fish and developed no diseases, as soon as Apex put sockeye into saltwater netpens they caught, according to Groves, "every disease in the book."[5] With sockeye eliminated as a commercially viable species in the shallow, poorly flushed waters of Genoa Bay, and chum commanding very low market prices, Apex expanded its hatchery output of coho.

By harvesting coho before they reached maturity, Apex was able to produce pan-sized salmon in less than two years. However, it soon discovered that in British Columbia there was a relatively small market for a fish of this size, and most of the demand was being filled by Idaho trout that could be raised more cheaply. There was, however, a market for much larger salmon, and to take advantage of it, Apex switched to raising chinook, which in the wild can grow to 30 kilograms. By 1981 annual production had climbed to $2 million, and Apex negotiated an Enterprise Development Grant with the federal government, which would give it an infusion of $500,000 for improvements. In order for the company to obtain the grant, the government required Apex's corporate backer, Union Carbide, to provide a written commitment of continuing financial support. The multinational first balked and then cut off all further financing for Apex, and in spring 1982 Apex declared bankruptcy.

Salmon farm licence #3 was granted in 1974 to Indian Arm Salmon Farm

Ltd, owned by Vancouver geological engineer Doug Goodbrand and two silent partners. They established a hatchery and grow-out farm seven kilometres north of Deep Cove on the west side of Indian Arm, the northern branch of Burrard Inlet, and began production with 25,000 coho eggs, which they grew to pan-size and marketed in the fall of 1975. Between 1975 and 1979 they increased their intake of eggs to roughly 300,000 per year, including some chinooks and rainbow trout, but even with BC Packers buying a ten percent share in the company, there was little financial gain for the partners. In the winter of 1979–80 they sold off their fish and closed the operation down.

In July 1976 Allan Meneely, no longer able to battle the combined forces of government bureaucracy and nature, ceased operations at Moccasin Valley. However, in spite of his failure, the concept of salmon farming on the Sunshine Coast was still very much alive, and by August 1978, John Slind of Suncoast Salmon Ltd. was granted salmon farm licence #4 for a hatchery and salmon farm on the east side of Sechelt Inlet between Tuwanek and Nine Mile points. The next spring, he and his brother, Stewart Slind, started a second site on the west side of Sechelt Inlet in a bay lying between Skaiakos and Piper points, wholesaling their first pan-sized coho that fall.

After watching a television documentary on the Japanese salmon farming industry, Percy Priest of Port Alberni established Alberni Marine Farms (licence #5) with his son, marine biologist Bill Priest. San Mateo Bay, their sheltered site on the south shore of the Alberni Canal opposite Tzartus Island, had deep water and good tidal flow, and the Priests successfully battled seals and sea lions, but they never solved the problem of freighters ploughing too fast up the inlet. The wake from one of them struck the cliffs behind the farm and, as it receded, tilted the netpens on their side, allowing 10,000 fish to escape. In 1978, when the Priests at last had their first fish ready for market, they hawked them from restaurant to restaurant in Vancouver and Victoria. Four years later, with their first major harvest, they signed up with a broker who flew the fish to eastern Canada every Monday morning. In the mid-1980s they sold the farm to the company that ultimately became International Aqua Foods.

Dan Gillis applied for salmon farm licence #6 in the name of the Nimpkish Indian Band of Alert Bay in 1977 as part of a larger project that included upgrading band members' basic education and giving them fish farming skills. Gillis and marine biologists Mike Berry and Bill Pennel set up a hatchery in an old floatplane hangar and brought in 25,000 coho eggs from DFO's Quinsam Hatchery. The survival rate was excellent, but only half of the fish smolted effectively before winter came, and the project had no facilities for overwintering the remainder. The project continued for two more years with 50,000 coho eggs each year, but it was not commercially viable and had to be shut down in 1980.

Mavis Smeal and her son Robert, who had worked at DFO's Capilano Hatchery, took out licence #9 in 1979 for their Saltstream Engineering Ltd., which they located on the family property at Doctor Bay on West Redonda Island. They built the hatchery and farm buildings from logs milled on site, then installed a pelton wheel to harness the creek water for electricity. Their first eggs—coho from DFO's Qualicum River hatchery—arrived on January 15, 1981, and their netpens were put into the water later that year. During the 1980s Saltstream was considered a model farm, and on one occasion five cabinet ministers came to inspect the operation. Of all the pioneer farms, Saltstream's story is the happiest: in 2003 the farm was still owned by the Smeals, and its netpens of all Pacific salmon stocks were still riding the waves of Doctor Bay.

West Coast Fishculture Ltd. (licence #11) was the pioneering project of Ward Griffioen, who came to BC after fish farming in the Netherlands. He chose Great Bay in Bedwell Sound, northeast of Tofino, for his hatchery/seapen operation.

By the end of the 1970s, salmon farming had been attempted at eight BC locations—Ocean Falls, the Sunshine Coast, the Nanaimo area, Barkley Sound, Indian Arm off Burrard Inlet, Alert Bay, Redonda Island and the Tofino area. The fact that so many of these projects failed, at least in their first incarnations, can be attributed to their isolation, seat-of-the-pants technology, failure to attract financing and lack of support from government. But all of these factors would change radically in the next five years.

Salmon Farming on the Sunshine Coast

Though the Meneely family had failed in their attempt to farm salmon in Agamemnon Channel, and the Slinds had not become fabulously rich on the profits from their Sechelt Inlet farms, by the early 1980s the waters of the Sunshine Coast had become the location of choice for prospective fish farmers. Jervis and Sechelt inlets—including the waters of Salmon and Narrows inlets—offered literally hundreds of farm sites; a mild climate assured comparatively safe working conditions on the water; and Lower Mainland markets beckoned no more than 90 minutes away.

With advice and encouragement from the scientists at PBS, DFO's

Capilano Hatchery and the Vancouver Public Aquarium, Brad and June Hope established Tidal Rush Farm on Nelson Island in 1978 and constructed a hatchery for their chum, chinook and coho eggs. They installed the first smolts in their netpens by 1980. But 18-fathom-deep Hidden Basin had a mud bottom, and during a prolonged hot spell that summer, hydrogen sulphide—a product of the decaying organic matter in the mud—bubbled upward, killing most of their first crop. Their second, however, was far more successful and by the spring of 1981 they were selling their fish in Vancouver and Calgary. Most of these salmon were in the 200- to 900-gram category, but the Hopes were also test-marketing salmon up to two kilograms.

In 1980 Tidal Rush received the first Ministry of Agriculture and Food (MAF) loan guarantee for aquaculture in BC. Further guarantees were awarded to the Hopes in 1982 and 1984, and their farm was designated as one of the sites for PBS's chinook strain comparison program, which was funded by DFO and the BC Research Council. The study showed that stock from the Big Qualicum River gave the best results in netpen farming, but it also revealed that the west coast of Vancouver Island—not the Sunshine Coast—was the most favourable growing area because the range of water temperature there was closer to that in prime Pacific salmon habitat. In spite of this information, in 1983 the Hopes took out a second aquaculture licence for a site just twenty kilometres to the northeast in Hotham Sound off Jervis Inlet and established another farm with its own hatchery, netpens and processing plant.

On April 19, 1985, the Hopes merged their Tidal Rush Marine Farms with Aqua Foods Ltd. (an oyster producer) and Deluxe Seafood Ltd. (a shellfish processing and marketing concern) to form Pacific Aqua Foods Ltd. (PAF), which then went public on the Vancouver Stock Exchange. PAF's prospectus claimed 41 netpens at two sites, a processing plant at Lund and a hatchery.

Just south of the Hopes' Hidden Basin farm, Tom and Linda May founded Cockburn Bay Seafarms in 1980. They put their first netpen into the water in 1981 and stocked it with 10,000 chum salmon purchased from the Hopes. All but 27 became dinner for blue herons. With their next crop, after many trials, they discovered that they could outsmart the birds by increasing the height of the 30-centimetre fence surrounding their pen by just another 30 centimetres and adding a line of electric wire around the top. Within two years they had established chinook broodstock to spawn their own egg supply and developed an annual capacity of two million smolts. By 1984, with their Cockburn Bay site at full capacity, the Mays made plans to build a hatchery on

Chapman Creek, which enters Georgia Strait five kilometres southeast of the town of Sechelt. When the Mays' marriage ended in 1986, they sold their Cockburn Bay Seafarms and Chapman Creek Hatchery to Dr. Sergio Kumar, an Italian economist and accountant, for $1 million. Kumar merged Cockburn Bay with his own company, Pacific Crystal Seafarms, and renamed it Royal Pacific Sea Farms Ltd. (RoyPac).

By this time, the Sunshine Coast Regional District (SCRD) had established an economic development commission to encourage investment in the area, and Oddvin Vedo, the commission's new director, took this to mean investment in the fledgling aquaculture industry. When Canadian bankers showed no enthusiasm, Vedo, who had come from Norway, turned his attention there, knowing that Norwegian aquaculture interests would leap at the opportunity to avoid the restrictions imposed by their own government.

In the early 1970s after disease, overproduction and saturated markets had crippled Norway's aquaculture industry, that country's government had passed an act "to lead and control the industry." It limited farm size and fish numbers to control disease and also imposed a moratorium on new farms while fresh overseas markets were developed. That ban was lifted in 1981, but there were still tight controls on farm capacities and ceilings on company expansions. Thus, BC's fledgling industry, with its lack of regulation, was an obvious place for Norwegian aquaculture companies to invest their profits.

While only a few years earlier Norwegian investment would have been determinedly discouraged by both provincial and federal governments, Canada's political climate was changing radically. The Conservatives had come to power in Ottawa in September 1984 and by the following summer had replaced the Foreign Investment Review Act with the Investment Canada Act, which was far more encouraging to foreign investors. Victoria followed the senior government's lead. Vedo, who in 1982 had led a group of eight BC salmon farmers on a provincially funded tour of the Norwegian industry, now worked with the provincial Ministry of Small Business and Industry to bring a Norwegian delegation to the coast to see the potential for investment. Fritjof Wiese-Hansen of Mowi AS, Norway's most successful fish farm, also visited the Sunshine Coast to scout farm locations. He had already acquired the major interest in a fish farming operation headquartered in Port Townsend, Washington, which he renamed the Scandinavian-American Corporation (Scanam).

Although there were only six fish farms on the Sunshine Coast by the summer of 1984, Vedo encouraged their owners to form the Sunshine Coast

Aquaculture Association. That September, two weeks after its inaugural meeting, the new association and the SCRD's economic development commission collaborated to hold the first Sunshine Coast International Aquaculture Conference.

A combination of the depressed economic climate, the stimulus provided by that first aquaculture conference, the widely publicized success of the Hopes and the Mays, and the apparent interest of Norwegian fish farming concerns prompted a flood of salmon farm licence applications for Sunshine Coast sites. In 1985 eight farms appeared in Agamemnon Channel, another in St. Vincent's Bay in Hotham Sound and yet another in Sechelt Inlet. Seastar Resource Corp., which had bought the Meneelys' old Moccasin Bay site, was the only major player among these new farm companies. Fronted by 37-year-old Italian businessman Franco Cecconi, the company had gone public on the Vancouver Stock Exchange (VSE) in November 1984, months before its first farm site was purchased. It was managed by Ward Griffioen, and technical advisers were former PBS colleagues Roly Brett and Bill Kennedy, now both retired.

Most of these new farm companies were launched with an average capitalization of $150,000, with some if not all of it coming from Norwegian banks such as Bergen Bartz or Christiana or from Norwegian companies such as ATI. Seastar Resource, however, had raised $500,000 on the VSE before receiving an infusion of $2.5 million in November 1985 from Royal Seafoods AS of Norway.

The BC government, spurred by these offshore companies' confidence in the industry, announced an Aquaculture Incentive Program (AIP) to provide fish farmers with five-year interest-free loans of up to $100,000 or 50 percent of eligible capital expenditure costs. The provincial Export Development Corporation agreed to insure up to 90 percent of export receivables for aquaculture companies, while the Ministry of Agriculture, under its Agricultural Credit Act, renewed its offer to guarantee 25 percent of salmon farm bank loans at one percent over prime rate. This program had actually been available to aquaculture since 1979, but Canadian banks had been reluctant to use it.

Unfortunately, what farmers needed most was operating funds, and neither AIP loans nor Agricultural Credit Act guarantees were set up to provide money for feed or wages. Canadian investors from the private sector also avoided funding operating costs, even though they had once again been assured of a 30 percent provincial tax credit for investing in BC aquaculture. Since few farms had yet shown a profit, operating costs were regarded as a bottomless pit. As well, environmental concerns about salmon farming were beginning to get media attention.

Diseases and Early Environmental Concerns

By September 1985 there were seventeen salmon farms operating on the Sunshine Coast, with a projection of 45 more farms to be established there within three years, and while the community generally welcomed the employment opportunities and the economic infusion, voices hostile to aquaculture were becoming louder and more insistent. As early as February 1985, Geoff Meggs of the United Fishermen and Allied Workers Union (UFAWU) speaking in Sechelt at a public forum on aquaculture declared the new industry "a hoax, a sea-bubble promise,"[6] that was destined to contaminate the wild fish. He said that DFO hatchery egg stocks intended to build up the wild resource were being used to stock fish farms, and he accused aquaculturists of trying to privatize wild fish stocks.

Fish Farm Wastes

Community organizations were also raising the first alarms that massive buildups of excess fish feed and feces collecting under the farms would contaminate the water and destroy the natural sea life. Many farmers were undoubtedly overfeeding, although certainly not by design, for while there was ample research to show that it was possible for a salmon to increase in weight by one kilogram for every two kilograms of feed ingested, few studies had been done to determine the correct rate of feed delivery to get that result. Fish farmers also had difficulty estimating the nutritional content of feeds available from different sources, so in spite of the price of feed (which accounted for half of a farm's operating costs), they erred on the generous side at feeding time.

To add to the problem, virtually nothing was known about the effect a large-scale aquaculture industry would have on small bodies of water like Sechelt Inlet, and cutbacks in DFO funding meant no testing was planned to determine whether salmon farming was already changing water quality there. However, local on-site inspections were revealing changes. In two years of regular diving, the manager of fish farms in Jervis and Sechelt inlets had watched the seabed below the farms turn into a wasteland covered with a film of fish food and feces, areas so dead that not even sea cucumbers or urchins inhabited them any more.[7] Tests done previously in Scotland under netpens with high fish densities had shown that feces and wasted feed that sank to the bottom resulted in eutrophication, which depleted the oxygen in the water, while natural decay processes created hydrogen sulphide and methane. Both caused the death of marine animal life. The process also released large amounts of nitrogen, which in turn increased the growth of phytoplankton, the tiny plants that drift or float in the water.[8]

In the natural marine environment, plankton are the staple of the food chain. Salmon and other large fish eat smaller fish, which feed upon even smaller fish, which feed on the drifting microscopic animals known as zooplankton. These animals graze on phytoplankton. There are literally thousands of species of phytoplankton in the oceans, but they fit into two main classes: the non-swimming, silica-shelled diatoms, and the flagellates that have whip-like appendages capable of producing minute currents that direct nutrients in their direction. Like land plants, phytoplankton require sunlight in order to grow and reproduce, so they are found in abundance only in the upper layers of the ocean. Here they multiply or "bloom" by using the sun's energy to turn simple nutrient salts such as phosphates and nitrates into tissue. Under prime conditions they can be found by the countless billions in this upper layer of water.

A study of the phytoplankton ecology of Georgia Strait published in 1979 states that "standing stocks" of plankton in that waterway had increased substantially since the end of the 1950s. The report blamed Vancouver's sewage. Traditionally discharged into Burrard Inlet, by 1974 it was being diverted into the Fraser River, resulting in increased concentrations of nitrogen and ammonia in the river. Based on observations made at sixteen stations in the strait between 1975 and 1977, the authors showed that annual phytoplankton growth had increased substantially where the river water, laced with land drainage water and sewage, swept out into the salt water. By mid-spring these increases were developing into blooms in the sheltered and protected waters of the southern strait, and by late May or early June were blooming more generally in its northern reaches.[9]

Since all animal life in the sea is ultimately dependent on the existence of phytoplankton, most increases are beneficial, although there are recorded cases of unusually large plankton blooms causing the death of marine animals when the plankton began dying and decomposing, lowering the oxygen content of the water.

Harmful Algal Blooms

Besides these generally beneficial plankton, however, there are a number of phytoplankton species that are inherently harmful and cause illness or death to other marine creatures and/or humans whenever they bloom. These harmful algal blooms (HABs) were recognized initially in relation to shellfish and are known to have claimed human victims on this coast as early as 1793.

But it was the introduction of netpen salmon farming to the Pacific Northwest coast that prodded scientists to begin studying a whole new range of HABs in these waters. While their existence had been suspected as early as 1942, when salmon held in traps off Sooke had mysteriously died, it was only in 1961, after the death of netpen-reared fish in Washington state, that the diatom *Chaetoceros convolutus* and the closely related species *C. concavicornis* were pinpointed as the cause. From Nanaimo, the PBS reported losses from *C. convolutus* and "related species" in three of the four years between 1973, when the experimental farm opened, and 1976.

Both species of *Chaetoceros* have small barbs on their setae that allow them to collect on the salmon's gill tissue. It was at first believed that the barbs penetrated the tissue, causing the fish to die of capillary bleeding, but it is now thought that the barbs irritate the tissues, causing a massive discharge of mucus; this obstructs the gill passages so that the fish suffocate as their oxygen supply is cut off and carbon dioxide builds up in their blood.

Fortunately for salmon farmers in the Pacific Northwest, *Chaetoceros* does not usually form heavy blooms in these waters, being instead a chronic, low-level member of the plankton community. When blooms do appear, the main concentration is not at the surface, but five to ten metres below it, with substantial amounts as deep as twenty metres. The scientists at PBS therefore recommended that in the event of a *Chaetoceros* bloom, salmon farmers should cut off feed so their fish would have no reason to rise into the range of the bloom. Unfortunately, the pens on most Sunshine Coast farms were not deep enough for the fish to get below the danger zone.

Equally threatening to the aquaculture industry was the presence of the harmful flagellate algae *Heterosigma carterae,* which has formed blooms in Pacific Northwest waters every year since the first observations of it were made in the 1960s. Until 1985, *Heterosigma* blooms in Sunshine Coast waters had all been minor, and no serious damage had been done to fish stocks, but the burning question was whether blooms were increasing there and whether salmon farm pollution was causing the increase. Since none of the sampling stations in the 1979 phytoplankton ecology study had been located in the waters of Jervis Inlet or Sechelt Inlet, government agencies had nothing with which to compare the present bloom frequency. In 1985, however, most BC salmon farmers were "totally unaware of the potential impact of harmful algal blooms," according to Dr. F.J.R. "Max" Taylor of UBC's oceanography department. "There was a lack of realization that these organisms are found right around the world within certain temperature ranges. Much of the reason that southern Norwegian waters are not used for fish farming is because of the HABs, and the Norwegians weren't being totally frank to BC farmers about the dangers."[10]

Bacterial Diseases

Most salmon farmers were concentrating their attention on the ongoing damage caused by bacterial kidney disease (BKD) and vibriosis, for which no effective controls were then known. Although the bacteria that cause these diseases can grow when the water temperature registers as low as 5°C, the optimum temperature for the BKD bacteria, *Renibacterium salmoninarum,* is 15°C, and for the vibriosis bacteria, *Vibrio anguillarum,* between 18 and 20°C. These temperatures are within the summer range of the shallower areas of Sunshine Coast and southern Georgia Strait waters. Both bacteria were probably present in Pacific Northwest waters long before salmon culture was established here, but by the mid-1980s concentrations appeared to be increasing in the neighbourhood of salmon farms.

First reported in Scotland in 1933, BKD has subsequently been found in virtually all countries in the northern hemisphere where salmonids are farmed. Internal evidence of the disease may include greyish abscesses in the organs. Externally, affected fish are generally lethargic, dark coloured and partially blind with protruding eyeballs. Many have abscesses under the skin, which break open into the water. By 1984 studies had proved that it was these abscesses, as well as the feces of infected fish, that spread the bacteria. It was later demonstrated that infection was also passed from spawning females to their eggs. Attempts to control the disease with drugs or chemicals mixed into fish feed had so far proved of little use, and once fish stocks had become infected, slow and continuous mortality rates could account for as much as 50 percent of the population. Spontaneous outbreaks of BKD among wild fish were unusual, probably because of the low concentration of the bacteria in deep seawater, but transmission from netpen stocks to wild stocks was already known to occur.

At least 48 different species of fish are known to be susceptible to *Vibrio anguillarum* infection. Salmon affected by vibriosis exhibit reddening around the mouth, gill coverings, anus and base of the fins. Boil-like lesions grow under the skin and in muscle tissue, many of them breaking to the surface to become open sores. Internally, the fish have tiny hemorrhages in the visceral organs and peritoneum. By 1985 researchers were conjecturing that bacteria were distributed via feces and from the carcasses of dead fish. Salmon in overcrowded netpens with reduced dissolved oxygen and high levels of excretory matter in the water were found to be most susceptible to vibriosis. The recommended treatment for infected fish stocks at that time was to add sulfonamides and antibiotics to their feed; inevitably, some of these chemicals were transferred to the water surrounding the netpens.

As the general public learned of the existence of BKD and vibriosis, they also learned that there were no regulations in place to monitor them or the drugs used to combat them. The scientists at PBS recommended the correct drug dosages to the farmers, but these amounts were actually dictated by Agriculture Canada and the Food and Drug Administration, as they were for any other agricultural business. However, neither PBS nor Agriculture Canada made inspections, and DFO was responsible only for drug dosages at hatcheries, not at fish farms.

Hormone Contamination

Hormone contamination of local waters was also a concern to the community. By 1985 researchers were developing methods of feminizing salmon to deal with "jacking," a natural phenomenon in which a certain percentage of the males in any stock of Pacific salmon will mature, spawn and die prematurely, usually after only one year in the sea. In some strains of Pacific salmon these jacks cannot be distinguished from the others in the netpen by size, shape or colour until they begin to darken and lose their silvery brightness, by which time they have lost their market value. In other strains the jacks will be larger than the average fish in the pen, but so will the faster growers among the later maturing fish. If all the larger fish are marketed after one year in order to avoid taking a loss on the jacks, the best of the potential broodstock in the later maturing fish in the pen will also have been culled.

Since jacking is restricted to males in Pacific salmon, government scientists devised a hormone bath for newly hatched fish that would result in an all-female population, even though the former males remained genetically males. To reinforce the process, as the fish matured they were given hormone-enhanced feed. Beginning in 1984, a number of private hatcheries on the Sunshine Coast began testing these sex-control compounds under DFO supervision. Local residents protested that neither DFO nor any other government agency was conducting tests for water contamination at the hatcheries where these research projects were being carried out, nor were any precautions taken to prevent contamination of seawater with the hormone-enhanced feed.

Atlantic Salmon, Furunculosis and Sea Lice

The potential for new problems was developing in the wake of a 1984 DFO decision to permit Brad Hope of Tidal Rush Farms, IBEC Inc. (which had installed a farm on Vancouver Island) and PBS at Nanaimo to import Atlantic salmon eggs to their farm sites. Similar to Pacific salmon in taste, colour and texture, Atlantics were considered more desirable because even though they spent as much as two years in fresh water and were reputed to be notoriously difficult to culture in the fry stage and overly sensitive to environmental stress, they grew twice as fast as chinook or coho once they were put into saltwater pens. In addition, Atlantics could be brought to maturity and marketed when wild Pacific salmon were not available, and they commanded a significantly higher price in the US market due to a successful Norwegian advertising campaign.

The importation of fish eggs was at that time governed by the federal Fish Health Protection Act, which stipulated that eggs, fry or smolts must come from a facility that used precisely defined methods and met exacting maintenance and operation standards. When a hatchery in Scotland passed the scrutiny of DFO's inspectors, an order was placed for 130,000 eggs to be shared by the three BC facilities that had been granted import permits.

In the mid-1980s Norway was producing 26,000 tonnes of farmed Atlantic salmon annually while Scotland produced another 8,000 tonnes, but by then the disease furunculosis was causing annual mortality rates of 30 to 40 percent in both countries. Although identified in Scotland over a hundred years earlier, furunculosis had not been a problem in Norway until the mid-20th century, when rainbow trout farms and hatcheries became infected. In 1968, in an effort to eradicate the disease, the Norwegian government had moved to stop imports of infected eggs and smolts. Its Diseases of Fish Act also provided for regular inspections of farms, full virological and bacteriological examinations of fish and compulsory reporting of mortalities and abnormalities. Four years later, aquaculture officials claimed that by rigorous inspections, steam cleansing and disinfection of hatchery facilities, and several months' fallow period between crops, all trout farms were disease-free. However, furunculosis had already been passed to one of the country's primary salmon rivers, and it soon became a source of infection for salmon farms. Stocks were reinfected in 1985 when diseased smolts imported from Scotland slipped past Norwegian inspection teams.

Furunculosis is caused by the bacterium *Aeromonas salmonicida,* which produces an enzyme that inhibits immunity-producing cells. The name of the disease comes from the furuncle-like lesions—red, purple or bluish in colour—that frequently appear on the fish's skin. They are caused by the breakdown of blood-vessel walls, which allows blood and serum to leak into the surrounding tissue. Diseased fish also show frayed or bloodshot dorsal fins. Inflamed spots are frequently found inside the body cavity, the kidney is surrounded by bloody fluid, and the spleen is swollen and changed from dark red to bright cherry red. Outbreaks of the disease generally follow the period in late summer when seawater temperatures have been holding consistently above 10°C. Fish that do not die of acute infection can become carriers and harbour the pathogen within their bodies, infecting other fish.

While furunculosis was becoming established as an endemic disease in Scotland and Norway, viral hemorrhagic septicemia (VHS) was causing annual losses exceeding US$40 million for Atlantic salmon farmers in the Scandinavian countries. VHS is a systemic viral disease that was first identified in 1949. It occurs primarily in cold water, with most outbreaks occurring when water temperatures are below 15°C, so mortalities are highest in the winter months. Infected fish usually have swollen kidneys and hemorrhaging in the body cavity; external symptoms include "pop eyes," darkened skin and lethargy. Since no medications or vaccinations had been found effective, control was mainly through the destruction of infected stocks and disinfection of equipment.

Atlantic salmon were also peculiarly susceptible to "sea lice," the common name for the parasitic copepods *Lepeophtheirus* and *Caligus,* worm-like creatures up to 50 millimetres long and six millimetres thick that survive by attaching to a fish and consuming mucous or skin. Most wild salmon play host to small numbers of them, but sea lice infestations occur in unnaturally high densities in salmon farms due to the large numbers of fish packed into a small area. An infestation can be disastrous. As many as a hundred of the parasites will attach themselves to a single fish and eat its flesh away. Infestations are generally confined to the warmer months of the year; in the winter months the lice tend to drop off the fish while new adult lice begin slowly developing from the egg stage. Aquaculturists had had limited success controlling outbreaks of this parasite in Norway and Scotland by pouring biocides directly into the water, but losses still ranged up to 50 percent in infested pens.

The perception that the introduction of Atlantic salmon brought heightened risk of disease to British Columbia waters sparked a fresh outcry from area residents and fishermen. They also worried about the consequences if imported fish escaped from their pens and interbred with the already endangered wild Pacific salmon stocks. However, the farmed salmon industry's spokespeople and government officials issued reassurances that the imported eggs came from Scottish hatcheries that had been guaranteed disease-free for seven years, that Atlantic salmon would not escape and that, in any case, they could not crossbreed with Pacific salmon in spawning streams. They pointed out that although 5.5 million Atlantic fry, fingerlings and smolts had been introduced into the Cowichan River between 1910 and 1930 in an attempt to establish an Atlantic salmon run for sport fishermen, the experiment had been completely unsuccessful. Skeptics, in turn, reminded the spokespeople that brown trout, which are closely related to Atlantic salmon, had been introduced to the Cowichan and Little Qualicum rivers during that same period, and this species had survived and expanded its population, proving that the introduction of any exotic species into a new environment is irreversible, and the ultimate consequences unknowable. However, the program of egg importation continued.

Public Resentment Boils Over

As government ignored local questions and protests, people on the Sunshine Coast began to believe that the salmon farming industry had been given carte blanche to defile the landscape and take over waters that had traditionally been available for recreational use. Public resentment grew, and in the summer of 1985 it was released against Scantech Resources Ltd., a company that had been formed early that year by three Sunshine Coast men—Carsten Hagen, Clarke Hamilton and Bernt Rindt—with the backing of the Norwegian firm Jamek AS. As well as processing and marketing fish, the company planned to provide prospective fish farmers with a "turnkey" service that included selecting farm sites, obtaining all necessary licences, arranging contracts for supplies, installing equipment and providing staff.

Scantech planned a processing plant at Earl's Cove on two waterfront hectares owned by Scantech shareholder Bernt Rindt. The property zoning allowed for aquaculture, but in order to obtain a building permit for the plant, the company had to apply to the Sunshine Coast Regional District (SCRD) for a bylaw amendment that would change the definition of the word "aquaculture" to include the processing of fish. Residents of the tiny hamlet of Egmont, close to the proposed plant, protested the impending noise and odour pollution, the possibility of water pollution from bleeding and gutting the fish, and the deterioration of the only road into the community as a result of the extra traffic going to and from the plant. In response, the SCRD called a public hearing for June. The company ultimately won the right to build the plant, but long before that happened it had already enraged another Sunshine Coast community.

This second controversy erupted on Sunday, July 14, 1985, when Scantech towed its first turnkey operation into place on the foreshore of property belonging to a minority shareholder, Andrew Sterloff. The property's zoning permitted aquaculture, but it was situated on Wood Bay, adjacent to a long-established, upper-income residential area, and as soon as the possibility of a fish farm at this site was noised around, the residents had begun protesting the potential visual pollution as well as the contamination of local waters. Ignoring the protests, Scantech staff continued acquiring the necessary permits and licences for the new Wood Bay Salmon Farms, but three days before the planned installation they discovered that they had overlooked the most important one: the foreshore lease. Belatedly, they sent the application via the usual route through the Burnaby office of the Ministry of Lands, Parks and Housing (MLPH). Normally, MLPH would then have sent copies of the application to other ministries, DFO and the SCRD to get comments. However, when it became apparent that there could be months of delay in the case of the Wood Bay application, as well as the possibility of a public hearing on the suitability of the site, Scantech's president appealed directly to Tom Lee, the minister of Lands, Parks and Housing, stating that 175,000 fish waiting at Cockburn Bay hatchery to go into the Wood Bay pens would die unless immediately transferred to the pens at their permanent site. Lee phoned the Burnaby office and directed staff to "fast-track" the licence. Wood Bay neighbours awoke the next morning to find the fish farm on their waterfront doorstep, and although the company's licence had been classified as temporary, Clarke Hamilton assured Wood Bay resident Mac Richardson that the farm would be there "for a long time."[11]

When Wood Bay Salmon Farms' new managing partner, Roger Engeset, arrived from Norway to take over the farm, he walked into a tempest. The storm clouds grew even blacker in September when local residents protested that "a cloudy layer of fish effluent which later clumped together into brown patches on the surface of the water" was coming from two pens of smolts at the farm.[12] There was no investigation by government.

The MLPH finally referred the Wood Bay application to the SCRD for comments in October, but that body's response was still pending in December when Scantech added another six seapens to the Wood Bay project. Tempers flared higher, even though Scantech was able to prove that the pens had always been part of the approved development plan for the farm. In May 1986 Scantech applied for permission to build a pipe out into the bay "to facilitate blood-water disposal"; the SCRD refused permission on the grounds that the company had promised there would be no fish processing at the site, but Scantech had already begun construction of a processing plant.[13]

The Industry Expands

Although the aquaculture industry's primary focus the first half of the 1980s was the Sunshine Coast, salmon farms were being established with little controversy and only passing media attention in other coastal areas. Most of the sites chosen were in the general area of the farms pioneered in the late 1970s and early 1980s: Cortes, Quadra and Read islands at the north end of Georgia Strait, Homfray Channel off Desolation Sound, and Sansum Narrows and Nanoose Harbour on Vancouver Island's east coast.

As on the Sunshine Coast, buyouts and amalgamations marked the progress of fish farming everywhere on the BC coast during the mid-1980s. Quality Seafarms had scarcely towed grow-out pens into Read Island's Little Bear Bay in the spring of 1985 when the company was bought out by Norwegian Enterprises Inc., generally known as Norent. One of the last BC salmon farms financed directly by the Christiana Bank, Norent had been

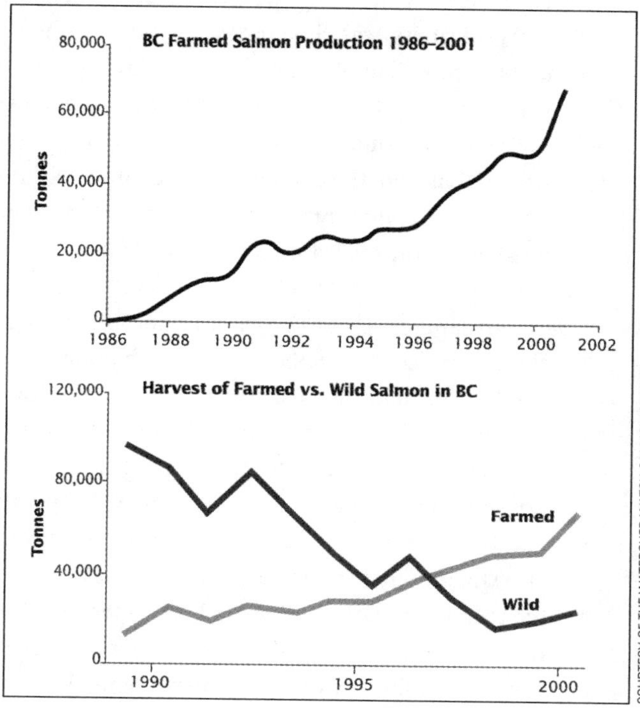

BC Farmed Salmon Production 1986–2001
Tonnes
80,000
60,000
40,000
20,000
0
1986 1988 1990 1992 1994 1996 1998 2000 2002

Harvest of Farmed vs. Wild Salmon in BC
Tonnes
120,000
80,000
40,000
0
Farmed
Wild
1990 1995 2000

COURTESY OF THE WATERSHED WATCH SALMON SOCIETY

organized by four Norwegians, led by managing director Kjell Aasen. By 1986 Norent was farming Atlantic salmon exclusively, bringing in eggs from the Altmore Fish Hatchery in Scotland.

The first salmon farm in Tofino Inlet was owned by Sea-1 Aquafarms Ltd. The company soon bought a second site, applied for licences for two more sites in the area, began construction of a hatchery and purchased Alberni Marine Farms Ltd., the Priest family's pioneer farm at San Mateo Bay in Barkley Sound. However, by April 1987 the company was deeply in debt and forced to reorganize. Under a new name, General Sea Harvest Corporation, it listed its shares on the stock market and began concentrating on joint ventures in salmon farming.

Western Harvest Seafarms Ltd., with three seapen sites in Esperanza Inlet north of Tofino, also turned to the stock market for funding in January 1988 and entered into a marketing agreement with BC Packers in exchange for operating funds. Other independent salmon farms adopted similar solutions to their marketing problems.

Sea Farm Canada Inc., based in Campbell River and with a first farm on East Cracroft Island, was a joint venture between Canada Packers Ltd. and Sea Farm Norway Inc. The latter was owned by Jacob Stolt-Nielsen Jr., whose primary business was Stolt Tankers and Terminals (Holdings) SA, a fleet of tanker ships hauling chemicals worldwide. He pioneered Atlantic salmon smolt production in Norway and by the 1980s had become the country's leading smolt producer. Since Sea Farm Canada's production was limited to Atlantic salmon, the company entered into a joint venture at the beginning of 1986 with Pacific Aqua Foods (PAF) to build a hatchery for Atlantics near the head of Neroutsos Inlet on northern Vancouver Island. This became the first major producer of Atlantic salmon smolts on this coast.

In January 1986 MLPH announced new regulations for aquaculture. A simplified licensing policy provided that anyone wishing to establish a farm would automatically receive a one-year "Section 10" MLPH lease of the required foreshore. The licensee would then be required to prepare a feasibility study and a development plan for approval by the Ministry of Environment in order to receive a ten-year operating licence. Technically, the approval of DFO, the agency responsible for protecting wild fish habitat, was also still required for the operating licence, but the MLPH could now override the recommendations of the federal fisheries department, the provincial environment ministry and local governments. Licensees were not supposed to begin development of the property while holding a Section 10 licence, but there was nothing to prohibit them from doing so. Responsibility for the actual farming of fish was transferred to the jurisdiction of the provincial Ministry of Agriculture and Fisheries (MAF).

With this change in the balance of power, the DFO's role was confined to what it could do through the Fisheries Act by prohibiting, for example, the deposit of "deleterious substances" into the water or disturbing or altering the environment. Up to 1995 there were no specific regulations in the act for the control of salmon farming "because," according to DFO spokesman Ron Ginetz, "we do not believe that salmon farming causes a deleterious impact on the environment. And if it does in certain instances, it can be dealt with through the existing act."[14] The DFO's mandate also allowed it to undertake research on disease identification and disease therapeutins and conduct environmental impact studies on subjects such as the effect of predation by farmed fish on juvenile wild fish, antibiotic residues in the water column around fish farms, and drug residues in farmed fish.

The provincial government's simplified regulations were a boon to prospective salmon farmers. Applications for farm sites on the Sunshine Coast flooded into MLPH's offices in spite of growing opposition from the community and a September 1985 warning from Dorothee Kieser of the Pacific Biological Station that summer temperatures in the waters of Georgia Strait as far north as Cortes Island could reach 20°C, well above the ideal 12°C to 15°C for raising salmon. Perhaps the farmers preferred the opinion of the BC government's aquaculture coordinator, Jim Fralick, who said that if the fish had survived the high temperatures of summer 1985, they had passed the acid test.

By June 1986 more than 50 applications had received approval; an additional 70 were pending. Of these, two had been submitted by Scantech for sites on the foreshore of Keats Island in Howe Sound. The islanders reacted by bombarding government and media with their opposition. Tom Siddon, federal minister of Fisheries and Oceans, tried to reassure them that it was in the self-interest of the aquaculturists "to make sure there was no major catastrophe,"[15] but to no avail.

Fifteen applications came from Aquarius Seafarms of Hotham Sound for sites in Sechelt Inlet. In addition, it asked that a piece of land near Scantech's Egmont processing plant be rezoned for its own processing plant. After delaying the rezoning, the SCRD petitioned the provincial government for a moratorium on fish farm licences, held a foreshore seminar in March 1986 to answer the public's questions, and struck a foreshore advisory committee to resolve water use problems. In late May the provincial government responded to public concern by announcing an aquaculture review for the Sunshine Coast to "analyze the opinion of local residents with regard to the establishment and development of the industry on the foreshore areas."[16] The ministry also promised to develop guidelines for improved public input into the decision-making process—even as MLPH continued to approve fish farm applications.

The only real setback for the industry came in July when the BC government announced the closure of Howe Sound to future aquaculture licences because of deteriorating water quality. Scantech's licence applications for sites off Keats Island were the only ones directly affected, and after the announcement Scantech withdrew them. Keats Islanders held a gigantic celebration, but it was a small victory for the forces opposing salmon farming because as the various levels of government postured and wrangled, the farms newly approved for Sunshine Coast waters were being towed into place and hastily stocked while other farms were being vigorously expanded.

The Bloom of 1986 and its Aftermath

On July 7, 1986, a bloom of the phytoplankton *Heterosigma carterae* enveloped the salmon farms of the Sunshine Coast, particularly those in Agamemnon Channel and Sechelt Inlet. The penned fish died when massive discharges of mucus obstructed their gills, preventing them from taking oxygen from the water. The farmers tried to save fish by pumping water from the lower depths into the pens, bringing in compressors to oxygenate the water, and covering the pens with black plastic to force out the light-loving plankton, but losses were very heavy (twelve farms reported 100,000 lost), especially since it was mostly mature fish that were affected. Many farms lost broodstock weighed down with eggs. Others tried to process and sell fish, but consumers feared they might be loaded with antibiotics. Sales of all farmed salmon suffered.

The directors of the SCRD, seeing the bloom as proof that there were too many fish farms, renewed their demands for a licence moratorium in Sechelt Inlet. The response of the minister of Lands was to release a handbook for prospective aquaculture investors. It was based on a Lands Branch preliminary draft of a Coastal Resources Inventory Study (CRIS) and included a map marked with red zones (where aquaculture would conflict with already established resource uses or fragile ecosystems), yellow zones (where fish farming would be moderately acceptable) and green zones (where fish farming would be highly acceptable to the public). Georgia Strait waters from Gibsons to Pender Harbour were marked in red; the north end of Sechelt Peninsula was yellow; and all of Sechelt, Narrows and Salmon inlets were green. There would be no moratorium, said the minister. Market conditions, not government, would determine the density of fish farms in the inlets.

However, some people in Victoria were convinced there were problems in the Sunshine Coast aquaculture industry. Two months after the *Heterosigma* bloom, mariculture biologist Ed Black, from the Ministry of Environment, issued a new warning of the potential for further pollution in Sechelt Inlet. He explained that only the top 55 metres of water circulates regularly, with the lower depths retaining their deposited sediments undisturbed. On the average farm, he said, for every 100 tonnes of feed distributed over the pens, twenty tonnes was going through the mesh and onto the seabed. When combined with fish feces, this excess feed could produce more than seven tonnes of nitrogen, of which 1.5 tonnes would be retained in the water, enhancing phytoplankton blooms. He theorized that the proposed maximum 4,500-tonne annual production of finfish in Sechelt Inlet could result in a bloom 18 metres deep, 90 metres wide and 3.7 kilometres in length.[17]

Critics compared Sechelt Inlet's problems to those of Japan's Inland Sea, where yellowtail and other species had suffered from harmful algal blooms (HABs) for many years before the source of the nutrients for the blooms had been identified as the extraordinarily high level of phosphates in the City of Osaka's sewage. By drastically cutting these levels, the incidence of HABs dropped by 50 percent, but because the sediments remained full of harmful algae, each time they were stirred up, the blooms recurred. It was predicted that Sechelt Inlet would follow the same scenario, but with fish farms as the source of nutrients for the blooms.

Dr. Max Taylor of UBC's oceanography department, internationally recognized for his expertise on plankton blooms, came to Sechelt to investigate where the bloom had originated. His team of researchers monitored plankton levels at nine locations in Sechelt Inlet from 1987 to 1989 and discovered that *Heterosigma* levels were always highest at the inlet's mouth where it meets Jervis Inlet. They concluded that the HABs had been formed in outside waters and swept inside by tides and currents. There was little correspondence, Taylor said, between Japan's Inland Sea problems and those of Sechelt Inlet: Japan's sea is very shallow, while the inlet has an average depth of 207 metres and, although almost completely enclosed, holds a very large body of water.

"The sediments here," stated Dr. Taylor, "have little effect on the water column; occasional overturns where deep water comes up near the surface only cause low oxygen levels in the upper water, not plankton blooms . . . There was no sign of the farms actually contributing to this bloom."[18] His researchers concluded, however, that the fish farmers in Agamemnon Channel and Sechelt Inlet had not seen the last of HABs because they were at the mercy of the tides and currents that brought them. In the aftermath of Taylor's report, only one farm was moved.

The *Heterosigma* bloom of July 1986 brought another festering problem into the public spotlight: the disposal of dead fish or morts. With thousands of dead, unmarketable fish on their hands, the farmers sought quick solutions, the quickest being to dig holes and bury them. Wood Bay received a permit from the Coast Garibaldi Health Unit to dispose of its morts in large pits; neighbours protested that the pits would attract bears and contaminate the groundwater. Other farmers buried their morts without benefit of permits, loaded them on barges and dumped them at sea, or trucked them to local garbage dumps where they rotted into soupy quagmires. When dump operators complained, restrictions were imposed limiting the amount of morts that could be accepted, the timing of mort deliveries, and arrangements for liming, but problems continued to mount, and the SCRD began enforcing the bylaw that made it illegal to bury fish wastes in pits. It also started enforcing the law against burning noxious waste in order to stop farmers burning plastic feed bags.

Meanwhile, in spite of July's disastrous *Heterosigma* bloom and Dr. Taylor's warnings, 400 delegates to the Third Annual Aquaculture International Conference in Sechelt in early September 1986 learned from Mike Coon, head of the finfish unit of MAF's Marine Resources Branch, that among the 32 operating salmon farms on the Sunshine Coast (compared with seventeen the previous year) there was still room for an additional 30 farms. Province-wide, the industry's statistics showed that the 70 operating salmon farms already represented a direct capital investment of $75 million, employed 375 people directly and annually spent approximately $50 million on services and supplies.

But while the industry was congratulating itself, the growing public outcry, plus advice from officials within the provincial environment ministry and pressure from the United Fishermen and Allied Workers Union (UFAWU), at last convinced Premier Bill Vander Zalm to place a temporary moratorium on all aquaculture foreshore leases as of November 1, 1986. A Kamloops lawyer,

David Gillespie, was appointed chairman of an inquiry into BC finfish aquaculture and given 30 days to come up with a report. The resulting document was presented to the provincial government on December 12, 1986.

Gillespie criticized all levels of government for the lack of a clear, comprehensive and integrated policy on aquaculture and recommended a federal-provincial agreement on the provision of services, setting of regulations and approval of licences. He said the provincial government should simplify its licence application requirements; show greater cooperation with local governments in promotion, planning and the approval of licences; set up a system of mandatory environmental monitoring for each aquaculture site and the surrounding area; and increase support for research into the long-range effects of salmon farming on the environment. He pointed out the damage done by the "gold rush" mentality that pervaded the industry and advised that the minimum distance between farms should be increased from 0.8 kilometres to three kilometres to reduce the impact on other resource users. On the subject of Atlantic salmon egg imports, he wrote that the risk of disease still "exists despite the severe screening, quarantine and effluent treatment standards imposed by government."[19] He called for the province to complete the CRIS of the Campbell River–Johnstone Strait area, the Islands Trust area and the Sunshine Coast, identified as the areas of highest user conflicts. He also recommended the immediate lifting of the moratorium on licence applications.

While the industry was understandably relieved, the UFAWU and most concerned citizens groups saw it as a step in the wrong direction, insisting that the moratorium should not be lifted until intensive, long-term environmental studies were completed and deploring the fact that the question of foreign ownership had not been addressed.

On January 27, 1987, the BC government announced that the moratorium on new licences would be lifted in stages beginning on April 30, by which time the studies of the Campbell River–Johnstone Strait area and the Sunshine Coast would be complete. Howe Sound and the areas of Georgia Strait administered by the Islands Trust would remain under study until June 30. The provincial and federal governments agreed to establish policies to regulate and assist the industry, and they signed a memorandum of understanding to identify and resolve barriers to aquaculture financing. Another joint agreement streamlined licensing procedures. Although the DFO continued to allow importation of eggs, fry and smolts into the Pacific region from certified foreign facilities, it imposed an added set of regulations stipulating that Atlantic salmon stock could only be imported as eggs; the eggs had to be surface-disinfected with an iodine compound before leaving the exporting country and again after arriving in BC; and eggs and the fish that hatched from them were then to go into quarantine on the importer's farm, with the quarantined fish tested every month for viruses.

The new regulations, while more restrictive, allowed the industry to continue the switch from chinook and coho production to Atlantic salmon, which were cheaper to raise. But obtaining high-quality Atlantic eggs was still a major problem. The only Scottish hatchery to receive Canadian certification before the more stringent rules of 1987 came into effect did not have the capacity to fill all the orders, yet, miraculously, all the orders were filled. It may be coincidental that one of the worst epidemics of furunculosis to hit the BC industry started in the farms taking delivery of eggs from that source that year. No Scottish egg imports have been approved since 1989.

BC hatcheries were also guilty of providing low quality and/or diseased stock of Pacific species. Bjorn Skei of Sechelt Salmon Farms entered into an agreement in 1987 to provide the Ocean Farms hatchery near Duncan with 3,000 11-kilogram chinook brood fish in the spring in exchange for 40,000 chinook smolts to be delivered to him that August. "Those fish never stopped dying," said Skei. "Within a year I had it in almost all of the pens, that same sickness. Two years later I was finished."[20] Nine other farms receiving smolts from the same hatchery were also affected. The disease was eventually diagnosed as marine anemia or plasmacydoid leukemia, an ailment that apparently affects only chinook salmon. Since it cannot be transmitted to humans, the DFO's Ron Ginetz advised the farmers to grow remaining stock to maturity and market them.

More Problems

In November 1986, during a regular inventory of its holdings, Pacific Aqua Foods (PAF) discovered that the actual number of fish in its Hidden Basin and Hotham Sound seapens was substantially lower than estimated. On January 20, 1987, the company announced that it had registered a $4 million loss for the six-month period ending November 30. PAF's stocks, which had peaked at over $8 in late August, closed at $3.50. Officially the losses were blamed on an inaccurate initial count of smolts, inaccurate record keeping and heavy losses to HABs and BKD; privately, insiders blamed the loss on poor site management. Brad Hope, who established one of the pioneer farms in 1978, was fired as company president.

Canadian banks and investment houses became wary once more, and even warier when *Chaetoceros* blooms swept into southern Jervis Inlet and the Campbell River–Quadra Island area in April 1987. However, investors from outside the country still saw BC as a prime location for salmon farms, and the industry continued to grow and mature, although at a slower pace. That September the Sunshine Coast Aquaculture Association held its biggest show to date, and by December a total of 152 farm sites had been approved, another 151 applications were pending, and 157 Section 10 licences had been issued for "site investigation."

By the end of 1988 there were seven salmon grow-out operations and two hatcheries in the Prince Rupert area. Royal Pacific Sea Farms (RoyPac), now the largest and most diversified salmon farming corporation in North America, had brought in two "super-farms" to raise Atlantic salmon (nearly 900 tonnes by 1989), anchoring them in Union Inlet and Worsfold Bay, both in Work Channel north of Prince Rupert, "just a stone's-throw-distance from the Alaska border."[21] RoyPac also owned six hatcheries, a processing plant and fourteen marine farms, among them three more super-farms. Although the company was listed on the VSE and TSE, Sergio Kumar remained its majority stockholder. After the company's first major harvest in mid-1988, projected sales for the entire year were set at $9 million, with $25 million in sales expected in 1989.

The outlook for the industry appeared to be promising. Mortality rates had dropped to approximately twenty percent, farmers were learning to harvest fish with early signs of disease before appearance and marketability were affected, and coastal weather patterns had stabilized.

There were a few clouds on the horizon, however, and a particularly gloomy one appeared to be hanging over Aquarius Seafarms Ltd. In spite of local protests the company had opened a processing plant at Egmont in 1988 and also owned twelve farms on the Sunshine Coast and a hatchery at Grey Creek on Sechelt Inlet. But behind the scenes things had begun to sour that fall when Ove Mjelde, board chairman, announced that, based on the initial six weeks of fall harvesting, previously forecast gross revenues for 1988–89 would be substantially reduced. Aquarius stocks dropped to $0.90 per share.

Morts were also causing problems. Although many farms disposed of morts at approved sites, incinerated them or had them ground up to be used as fertilizer, others were not so conscientious about establishing environmentally sound mort-disposal systems. Farms in remote locations often dumped morts into the ocean or on adjacent crown land. In August 1988, DFO foreshore biologist Rob Russell reported "two documented instances where pits were trenched to enter the marine environment, and a running mass of flesh and maggots was flowing across the beach."[22] As municipal landfill sites began to be closed to mort disposal, this illegal dumping increased. In one case, when the driver of a truck loaded with diseased fish carcasses was denied access to a landfill site, he simply dumped his cargo into a salmon-bearing stream.

On December 6, 1988, Aquarius, RoyPac, Seastar Resources and Westshore Seafarms were all charged with the illegal disposal of fish carcasses and feed sacks. Conservation officer Jamie Stephen described Aquarius's mort disposal areas on Nelson and Captain islands as a "rank, foul-smelling, seething mass of protein."[23] The excrement from the birds that had been drawn to the pit had killed all the undergrowth in the area. Aquarius was fined $5,000 for each of the two infractions. RoyPac's lawyer Michael Welsh succeeded in delaying sentencing, but Westshore Seafarms received a fine of $8,000 for "a pit overflowing with a seething mass of putrifying salmon carcasses" as well as "pancakes of plastic slag" from burnt fish feed bags on the beach near its St. Vincent's Bay site.[24] Stephen had been drawn to the site by the large number of eagles roosting nearby and found a pit two metres wide by fifteen metres long, just six metres above high tide, where bears were feeding on carcasses of fish that had died of BKD. Though the company was ordered to discontinue these unlawful practices, when Stephen returned in June 1988 the pit was overflowing. It was worse when he returned in October.

Fish escapes were also a growing problem because salmon farmers did not have the expertise to cope with net damage caused by marine animals or storms. One farm operator stated that up to a third of his coho stock regularly escaped between the time they entered salt water as smolts and when they were mature enough to be transferred to larger mesh pens. Operators also lost control of nets they were attempting to clean on-site, allowing large numbers of fish to escape.

In the summer of 1988, a five-person fact-finding mission sent to Norway by the UFAWU discovered that twenty percent of Norway's fish farms were on the verge of bankruptcy. They learned that the worst problem was an imported parasite called *Gyrodactylus salaris*, which had spread to dozens of streams. Norwegian scientists had successfully rid 28 rivers of the parasite by poisoning their waters with pesticide. Unfortunately, this cure also destroyed all other life in the rivers and in the saltwater beyond them.

The year 1988 came to a close with 80 companies operating 121 salmon farms in BC waters—up from ten farms only four years earlier. Victoria had approved another 53 farm licences and issued 157 permits to investigate potential sites. Still under consideration were another 147 applications for farm sites. A production-cost survey conducted for the DFO showed that costs ranged from $6 to $8 per kilogram of marketed salmon, depending on the size and efficiency of the farm. This broke down into 50 percent for debt, 44 percent for feed, 21 percent for processing and transport, thirteen percent for smolts, medicine and labour, eleven percent for management and insurance, seven percent for depreciation, and four percent for loan interests.

Privately, however, industry insiders were estimating that up to 75 percent of BC's salmon farms did not have access to enough operating funds to survive through 1989. Norway's economy had slipped into a disastrous downturn, and its banks, which already owned as much as 40 percent of the BC aquaculture industry, had virtually cut off further investment. Arne Nore of the Christiana Bank said, "Frankly, I don't know who will fill the gap. It would be impossible for Norwegian banks to come up with that kind of money."[25] His bank, in an effort to distance itself from salmon-farm financing, had established a venture capital firm to which it loaned money to handle these accounts, but the new company was as tight-fisted as the bank had become.

Canadian Bankers Association spokesman Al Droppo pointed out that since there had not yet been a full harvest of fish, BC's farms had not demonstrated that Pacific salmon could be raised for a profit. "We feel that it can but it's hard to lend until you have a little bit of hard evidence."[26] Further bank involvement was hindered because the Bank Act did not officially recognize the aquaculture industry, making it difficult for banks to accept fish farm assets as loan security. The banks also complained of a lack of clarity in provincial aquaculture policy. Although approximately one percent of all the money invested to date in the province's aquaculture industry had been provided by government, farmers complained that most of it had been spent on advertising. As a result, farmers in need of operating funds were turning increasingly to feed and equipment suppliers and fish processors for help. It made sense for feed companies to supply feed on credit to farmers in arrears so that their fish would be saleable and their outstanding accounts paid. It also made sense for fish processors to provide operating funds to salmon farmers in order to assure a year-round supply of quality fish.

The Crucial Year: 1989

The year 1989 began badly for BC's aquaculture industry and got steadily worse. After the death of Emperor Hirohito on January 7, tradition-minded

Japanese citizens stopped consuming salmon as a sign of respect. During this mourning period, Japan's salmon farms and wild fisheries continued to build up large frozen inventories, shutting BC farmed salmon out of that market for the entire year. Farmed salmon prices, which had fallen toward the end of 1988, remained low, approximately 20 to 25 percent down from the previous year. Farmers responded by allowing their fish to grow larger while they waited for better prices. However, time was running out for stocks infected with BKD as this disease is most virulent in late winter and early spring.

Then, on the night of January 31, the worst wind storm in 30 years struck southern BC. With temperatures as low as -15°C, for more than 80 hours the Alaskan Express, as it was nicknamed, roared down inlets at 185 kilometres an hour, whipping up three-metre waves. Farms in Jervis Inlet were hardest hit. Unable to reach shore, workers at Osgoode Creek Seafarms dressed in their wetsuits and rode out the storm on their careening 300-tonne, two-storey concrete barge. Nets were torn open and all the ready-to-harvest fish were released into the sea. A seven-metre aluminum boat was thrown up onto the barge, then swept off again and sunk. The company skiff was found eighteen kilometres away in Vancouver Bay. The eight pens across the inlet at BC Packers' Glacial Creek Seafarms broke free and sailed down the inlet. One of the five Aquarius farms in Hotham Sound was sunk and the superstructures of three more damaged. Winds from Bute Inlet buffeted the fish farms of Quadra, Read and Cortes islands, sheeting the walkways and buildings in layers of ice. Coulter Bay on Cortes and Evans Bay on Read both froze over, as did Cowichan Bay on Vancouver Island.

That one storm introduced as many as 100,000 farmed fish into the wild fishery; those that escaped from Osgoode Creek were infected with BKD. Less than a week later, commercial gillnetter Sonny Reid, fishing off Sechelt's Trail Islands, brought in a netful of Atlantic salmon. Publicity about his catch elicited dozens of similar complaints from commercial and sport fishermen, who feared these fish would introduce disease to wild fish and that genetically altered farmed fish would breed with wild stock. Ron Ginetz of the DFO's Aquaculture Division insisted that this was unlikely.

In March the hemorrhagic septicemia (VHS) virus was discovered in three wild coho at Neah Bay in Washington state and in one wild chinook at an Indian band hatchery in the San Juan Islands. Although none of the fish showed any clinical signs of the disease, the DFO began routine testing of both wild and farmed Pacific salmon in Canadian waters. Three months later the US government abruptly began enforcing its "Title 50" regulations, which

Salmon farm broken up by tides.

had been in place for twenty years but never used. This meant that as of July 1, all Canadian fresh or frozen farmed salmonid products could enter the US only if accompanied by a certificate verifying that the farm of origin was free of VHS and "whirling disease." The certification process, good for one year, required farm inspections and sampling of products.

Laboratory checks of hatchery smolts were showing a higher rate of positive results for BKD than in previous years, and warnings went out that 1989 would probably be the worst year yet for BKD in netpen stocks. Farmers adopted a wait-and-see attitude because it was not known whether positive lab results meant the fish would definitely get the disease.

The next problem arose on May 15 when the SCRD closed two landfill sites to future dumping of feed bags and gave fish farmers six months to find an alternate solution to the disposal of morts. Feed companies agreed to recycle their bags, but a solution for the mort problem was not as easy.

Paperwork also increased as new regulations for monitoring fish farm wastes went into effect on July 1. The Ministry of Environment's Waste Management Branch divided farms into three categories based on their annual feed consumption, and farmers were required to fill out a questionnaire every three months accounting for everything they put into the water: feed, antibiotics, pigments, anaesthetics, morts, blood from fish bled on site and domestic sewage. Larger farms were also required to report on monthly temperatures, dissolved oxygen and salinity profiles, hourly current speed, direction of sediment accumulation, ammonia and nitrate concentrations and benthic life. In addition, these larger farms were to carry out surveys of the major plants and animals found under the farm as well as adjacent control areas for contrast. Farms in shallow water were to set up transect lines beneath the pens to monitor sediment; farms in deeper water were to install suspended sediment traps. New regulations to "provide government with an auditing system and relief to purchasers of fish,"[27] required farmers to sign affidavits stating that marketed fish had not been treated with antibiotics within the previous 120 days. If treatment had occurred within that time, a chemical analysis of the flesh would be required. Offenders were to be fined.

By mid-June that year, diatom blooms had made water "soupy" in appearance in all of BC's fish farming areas. Most of these diatoms were not harmful, but both species of *Chaetoceros* were developing blooms in Desolation Sound and around Read Island, where fish losses became serious. Using joint funding from MAF and the aquaculture industry, the BC Salmon Farmers Association set up a "Phytoplankton Watch" 800-telephone number to keep its members informed.

As the months rolled on and BC's salmon farmers watched in dismay, world production levels for farmed salmon increased. The Norwegians, who had predicted production levels between 70,000 and 80,000 tonnes, actually produced close to 150,000 tonnes in 1989, three-quarters of the total world market demand. Chile, New Zealand and Scotland also had high production years, while in BC the industry predicted twice the tonnage of the previous year as the farms licensed since 1986 prepared to market their first crops. Canada's harvest of wild sockeye was also one of the best in decades, and by summer there was a glut of salmon on the world market. Prices registered a drop of 40 percent below the same month in 1988. It was now costing BC farmers $3 a pound to raise fish that they could not wholesale for more than $1.60 a pound.

Many of the early investors began dropping out of the industry. Most had been carrying heavy debt loads, making them extremely vulnerable to the market downturn. The weakest players were weeded out early, and on July 31 the Sechelt turnkey operation Scantech Resources Ltd. closed its doors. According to its president, Clarke Hamilton, "Scantech tried to do too much, to offer too many services and products."[28]

A series of harmful algal blooms plagued the coast from Burrard Inlet to

Desolation Sound through August and September. *Heterosigma* blooms were particularly dramatic on and around the Sunshine Coast, and many more companies went under as a result. Creditors were left with hundreds of thousands of dollars in unpaid accounts. Most farms never found buyers; some farmers simply sold off their remaining fish and abandoned their sites and equipment.

The most shocking announcement came on October 2 when Royal Pacific Sea Farms went into receivership, closing down fifteen salmon farms from Prince Rupert to the Sunshine Coast and putting 160 people out of work. Although RoyPac had sold 3,000 tonnes of salmon in 1989—more than in any previous year of operation—it was $15 million in debt.

Aquarius Seafarms found itself in the same predicament on July 3, 1990, when the company was unable to make its second-quarter interest payment to Aqua Technology and Investments (ATI), its major Norwegian debenture holder. ATI was then unable to make its own payment to the Bergen Bank, so it assigned its Aquarius debenture security and its 52.3 percent of Aquarius's shares to the bank, which in turn assigned them to Durango Enterprises Ltd., a Canadian subsidiary of Bergen Seafoods, which was also owned by the bank. Durango placed the company in receivership, leaving Sunshine Coast merchants with $260,000 in unpaid bills. A new company, Scanmar Ltd., was set up to operate Aquarius's sites until buyers could be found.

Market analysts put most of the blame for the industry's sad state at the end of 1989 on the price collapse, but some of the blame belonged to the farmers. Most had insufficient knowledge of algal blooms or disease symptoms and treatments. They had chosen unsuitable sites, which led to conflict with local communities, the wild fishery and predator animals. Many expanded prematurely without sufficient financing or research. With the exception of Royal Pacific, the farms going into receivership had all been installed between mid-1985 and 1988, so they were taking their first major harvest out of the water in 1989. Badly undercapitalized in the first place, they had accumulated crippling debt to get that far. With no new infusions of operating funds available, they had gambled on strong prices and a bumper crop in 1989 to balance the books.

As it turned out, the farms that were not unduly affected by storms, disease or HABs actually did have good harvests: 12,500 tonnes of salmon were produced on BC farms in 1989, an increase of more than 90 percent over the previous year. Unfortunately, the wild fishery once again harvested a bumper crop and so did the Norwegian salmon farming industry, bringing prices to an all-time low. They were pushed lower by distress selling and competitive marketing within the BC industry.

More farm companies might have survived the crisis year if their owners had not looked upon them as cash cows. After Seastar Resources declared bankruptcy, Ward Griffioen, who had left the company six months earlier, said, "[The owners] could have cared less if you raised salmon or bicycles in their pens. They were just into the game of wheeling and dealing in stocks. It took me two or three years to figure out that when they tell you how honest they are, they are cheating you right, left and centre. They built penthouses in Vancouver and just kept building up this incredible staff, and all they did was shuffle paper around so they could raise more money on the stock exchange. It was a real shame."[29]

The owners' attitudes affected farm workers as well. They had been so poorly paid and housed, and their working conditions had been so abysmal, that they felt little loyalty to their employers. Some were reported to have sat idle while fish died of disease or escaped in massive numbers. Reid Arnold, who pioneered salmon farming in Jervis Inlet, recalls seeing a whole pallet of fish feed floating down the inlet and hearing stories of Aquarius's workers dumping bags of feed off the barge rather than go out and feed the fish on cold, rainy days. Jon Van Arsdell, who taught aquaculture courses at Capilano College in Sechelt, remembers touring farms where dogfish swam inside the nets, feasting on salmon at their leisure.

Restructuring the Industry

Few industries in Canada have undergone the phenomenal short-term expansion that the BC salmon farming industry has experienced, then suffered such a disastrous crisis and survived to become a viable industry again. By the summer of 1990, although a third of the salmon farming companies that had been registered in 1988 had gone into receivership or changed hands, and many of the rest were in serious financial difficulties, there were still 72 firms operating 135 farm sites.

The failure of big companies like Royal Pacific and Aquarius, plus the host of smaller ones, now provided opportunities for the formation of new conglomerates as well as allowing the virtually painless expansion of existing operations. Takeovers and distress sales became the norm. Where sites had proved to be prone to HABs, only the assets found buyers; where sites were suitable but equipment poor, the assets of other companies were towed in to upgrade or replace them.

Two Norwegian banks—Bergen and Christiana—held most of the paper on the failing and fallen companies. Bankers like Arne Nore, vice-president of the Christiana Bank and a major shareholder in the bank's venture capital subsidiary, Christiana Marine, began acquiring properties as farmers defaulted on their bank loan payments.

On the Sunshine Coast, only 30 farms remained of the 56 operating there in 1988, and only eight of them were still independently owned. All were in financial difficulties. Most bankrupt farmers walked away empty-handed, and so did their creditors. A few farmers went into receivership, then bought back their assets with money from Christiana Marine. However, the prospect of increasingly destructive annual HABs convinced most farmers to leave the Sunshine Coast. By the time the Sunshine Coast Aquaculture Association announced that its 1991 convention and trade show would be held in Parksville on Vancouver Island, and the provincial government decided to close the Aquaculture Centre on the Sechelt campus of Capilano College, the focus of the industry had already moved north.

BC Packers Ltd. (BCP), one of the Weston International group, had invested in West Coast aquaculture in 1986 to ensure itself a year-round source of salmon, but the company's big move came in 1989 when it acquired a 25 percent interest in the former Apex hatchery at Westholme on Vancouver Island. In 1994 BCP purchased all the northern sites that had belonged to Hardy Seafarms and Aquarius. After restructuring and closing unproductive or HAB-prone sites, BCP owned three hatcheries, ten farms in the Broughton Island area and one at Newcombe Point in Salmon Inlet. The company switched to Atlantic salmon in 1993. Ted Needham, head of the company's aquaculture division, said that "the relative merits of Atlantics over chinook boil down to one simple comparison: in the time it takes a chinook to grow in the sea to 2.5 kg, the average Atlantic salmon has broken through the 5 kg mark."[30]

The wholly BC-owned company Pacific National Group Enterprises Ltd. also took advantage of the industry's failures to move into salmon farming after 1989 with four sites on Vancouver Island's west coast—three in the Bedwell Inlet–Cypress Bay area and one east of Meares Island—which had previously belonged to Royal Pacific and Seastar Resources. Pacific National built a processing plant at Tofino and by 1994 had acquired two more farms in Cypress Bay. The company specialized in chinook salmon and by 1995 was producing 42 percent of the world's supply of farmed chinook.

By the end of 1991 the first wave of collapsing companies and buy-ups was more or less complete. Many farms, however, were still teetering on the brink of bankruptcy and facing uncertainty, largely because of the situation in Norway. Threatened with countervailing duties for dumping salmon on the US market at less than cost, Norwegian salmon farmers had cut exports by half. To accomplish this, the Fish Farmer's Sales Organization (FOS), a monopoly set up to market all of Norway's farmed salmon, began freezing a quarter of the harvest, financed by a consortium of Norwegian banks and paid for by the farmers through a levy of 40 cents a pound on salmon sales. However, FOS failed to find new, non-traditional markets for the product, and the banks, in a crisis of their own, refused to extend more funding. In November 1991 FOS went under, hundreds of Norwegian farming companies faced bankruptcy and there was a fear that they would release 40,000 tonnes of frozen and 60,000 tonnes of fresh salmon on the world market, driving the price of salmon to new lows. At that point British Petroleum, the dominant company in the Scottish farmed salmon industry, undertook to find markets for the frozen salmon on condition that Norway's farmers cut production. The future of salmon farming began to look less desperate.

BC farmers, meanwhile, were not getting any breaks in weather conditions. Devastating wind and rain storms struck the west coast of Vancouver Island on the Remembrance Day weekend of 1991 and continued until the end of December, bringing nearly 2.5 metres of rain. Often 150 millimetres fell in a single day. In Esperanza Inlet, Western Harvest Seafarms' pens disintegrated and thousands of fish escaped. On December 19, three employees attempted to retrieve gear from one of the farms as arctic outflow winds roared down the inlet. Their boat was swamped and one man was drowned.

Disease was still wreaking havoc as well. Marine anemia or plasmacydoid leukemia had been diagnosed between 1988 and 1992 in about twenty locations on the Sunshine Coast, three in the Quadra Island area and two near Tofino. Afflicted farmed salmon became lethargic, anemic, "pop-eyed" and dark in colour. The DFO advised farmers not to breed any infected fish that survived the disease, but to rear them to marketable size and sell them off. Since the disease apparently affected only chinook, it was recommended that farmers switch to another species at the infected sites.

The parasite *Kudoa thyrsites,* which causes the condition known as "soft flesh" in farmed Atlantic salmon, began showing up in marketed Atlantics from BC farm sites by 1991. When this parasite has become established in muscle fibre, it produces proteolytic enzymes that break down the filaments of the surrounding muscle in order to provide nutrients for its own growth and the development of spores. The host fish does not detect the invasion of the parasite until it disturbs the membranous covering of the muscle fibre. Atlantic salmon, unlike Pacific fish species, have no host response to this parasite, and it continues growing until it ruptures, spilling spores and other material into the interstitial spaces between the muscle fibres. Affected fish deteriorate rapidly after harvesting and are unappealing in appearance and texture. By 1993 BC industry losses to "soft flesh" were estimated at $350,000, and the incidence of infection was increasing. There is no easy method for detecting *Kudoa* infection, as fish show no external signs that they have been attacked, so farmers of Atlantic salmon in BC developed the tactic of simply culling infected fish during processing and sending them to be canned.

New provincial regulations were established in January 1990 to govern the drugs and vaccines used to fight fish diseases. Except for the drug oxytetracycline, which had been approved by Ottawa's Bureau of Veterinary Drugs in July 1989 for non-prescription fish farm use, up to this time all therapeutants—including formalin, chloramine-T, salt, antibiotics, anaesthetics and hormones—had been available only when specifically prescribed by veterinarians. The new guidelines maintained the prescription-only status of these therapeutants, but also set strict limits on the amount of residue allowed in salmon products and on the time that must elapse between drug use and marketing of the product. Legally, the acceptable tolerance for any drug in fish flesh was now zero in Canada, but there continued to be great variability in the efficiency and reliability of the available tests.

HABs continued to plague the industry. Major *Heterosigma* blooms developed in Georgia Strait in 1990 and 1991, speeding the northward shift of the industry. They also occurred on the west coast of Vancouver Island in 1992, killing some 250,000 farmed salmon. Western Harvest Seafarms in Esperanza Inlet, already badly crippled by the storms of November and December 1991, lost nearly three-quarters of its stock. "We had to pump the morts out into the hold of a big seiner," said Sandy Brook, a former employee. "It was horrible. We tried to count them, but a lot of them were just skeletons by then."[31] After harvesting what was left, the company's Toronto-based owners sold most of the buildings to a sport fishermen's camp, while the farm at Saltery Bay in Esperanza Inlet was sold as an operating unit to a new Norwegian-owned company, Scandic Sea Food Ltd. Three years later Scandic was producing Atlantic salmon at three sites in that area.

By 1992 Dr. Max Taylor had completed studies of phytoplankton blooms in Sechelt Inlet and Barkley Sound and had proved that the blooms in Puget

Salming Farming Companies	*Ownership*
Creative Salmon Company Ltd.	**Unnamed Japanese investors Lions' Gate Fisheries Ltd.**
Grieg Seafood BC Ltd.	**Grieg Seafood AS, Norway**
Heritage Salmon Ltd.	**George Weston Limited**
Marine Harvest Canada/Nutreco Canada Inc.	**Nutreco Holding NV, Bergen, Norway**
Omega Pacfic Seafarms Inc.	**Pan Fish, ASA, Norway**
Omega Salmon Group Ltd.	
Pacific National Salmon/EWOS Aquaculture Ltd.	**Statkorn Holding, Norway**
Saltstream Engineering	
Stolt Sea Farm Ltd.	**Stolt-Nielsen SA, Norway**
Target Marine Products Ltd.	
Totem Oysters	
Yellow Island Aquaculture (1994) Ltd.	

Source: Ministry of Agriculture Food & Fisheries, Fisheries and Aquaculture Division:
Marine Salmon Farm Sites — June 2003
Tenures Included in the Salmon Aquaculture Policy Framework Cap, Dec 2001
BC Salmon Producers 2003

Subsidiary Companies	Farmsites
Tofino Aquafarms Ltd.	Tofino Inlet (4), McCaw Peninsula, Fortune Channel, Dawley Passage
	Hecate Channel (2), Cliff Cove
Connors Brothers Ltd. S.K.M. Enterprises Ltd.	San Mateo Bay (2), Simoom Sound, Broughton Inlet, Greenway Sound (2), Cypress Harbour, Macktush Bay, Raza Passage, Okisollo Channel, Raleigh Passage, Wells Passage Sonora Island
Hatfield Biotechnology Ltd., Kitasoo Aquafarms Ltd. (Kitasoo First Nation), Kyuquot Sound Farms Ltd.	Hoskyn Channel (2), Read Island, Nodales Channel, Amai Pt., Kanish Bay, Okisollo Channel, Sansum Narrows, Shelter Inlet, Calm Channel, Markale Passage, Jackson Passage (2), Captain Passage, Kyuquot Sound, Pinnace Channel, Quadra Island, Sonora Island, Mathieson Channel
	Barkley Sound
Seven Hills Aquafarm Ltd., Sonora Sea Farm Ltd.	West Thurlow Island (2), East Thurlow Island, Wishart Island, Goletas Channel (2), Chancellor Channel (2), Loughborough Inlet (2), Doyle Island, Goletas Channel, Sunderland Channel (2), Quatsino Sound, Richards Channel (2), Jervis Inlet, SE Frederick Arm, Homfray Channel, Cardero Channel, Port Hardy, Okisollo Channel, Queen Charlotte Strait, Samsun Narrows, Shelter Bay, Phillips Arm
ECWS Site Co. Ltd.	Clayoquot Sound (2), Saranac Island, Hecate Bay, Sooke Basin, Fortune Channel, Bedwell Sound (4), Ross Passage, Herbert Inlet (2), Millar Channel, Shelter Inlet (2), Sulphur Passage
	Doctor Bay, West Redonda Island
	Quatsino Sound (4), Gilford Island (2), Nodales Channel, East Thurlow Island (2), Baker Island, Broughton Island (2), Sargeant Passage, Midsummer Island (2), Swanson Island, Bonwick Island, Havannah Channel (2), Tribune Channel (3), Fife Sound, Doctor Islets, Upper Retreat Passage, Holberg Inlet, Indian Channel (2), Frederick Arm
Hardy Sea Farms Inc.	Salmon Inlet (3), Sechelt Inlet (3), Hardy Island (2)
	Jervis Inlet
	Discovery Passage

Sound, Georgia Strait and the open coasts of the Pacific Northwest were being distributed by winds and currents. He noted that there appeared to be an increase in the frequency and nature of the HABs in these waters. His most interesting revelations came, however, during regular sampling of water off Vancouver's Jericho Beach, when his group discovered that *Heterosigma* settles in a semi-dormant state in bottom sediments during the winter, rising to surface waters again only after the water hits 10°C, then multiplying rapidly as it reaches 15°C. They also found that destructive blooms of this phytoplankton only occurred off this beach if and when the salinity of the waters was diluted by freshwater at the rate of exactly fifteen parts per thousand, and that this only happened in June when the Fraser River reached the peak of its spring runoff. As long as these two conditions—temperature and salinity—remained stable at these levels, the nutrients brought by the river water would allow the *Heterosigma* population to explode and its mass would be thrust farther and farther out into Georgia Strait until portions broke off and were swept north. Carried by tides and currents they drifted into Jervis and other inlets on the way, although the bulk of blooms beginning off Jericho Beach ended up in Desolation Sound. The future health of farms in these areas, therefore, appeared to depend on the Fraser River. After 1991, because the Fraser's annual runoff was much lower than normal, salinity levels off Jericho Beach remained too high and nutrient levels too low for a major bloom to develop. As a result, only minor HAB episodes hit southern Georgia Strait between 1992 and 1995, and although destructive, they were not as ruinous as earlier blooms.[32]

Other researchers discovered that *Heterosigma* blooms also began off Nanaimo and were swept northward along the east coast of Vancouver Island. The source of blooms off the west coast of Vancouver Island was not determined, but scientists did establish that major blooms in Barkley Sound arrived there from the north. Taylor's studies, however, found signs that a resident population of *Heterosigma* in Alberni Inlet might be producing some of the pockets of bloom in the sound.

A new wave of salmon farming failures came in 1992 as the companies that had been teetering finally fell over the edge. The biggest splash came when IBEC announced bankruptcy in March of that year. IBEC's parent company, Arbor Acres of Connecticut (part of the British Booker Group), had provided high-quality equipment, but the company was, in the words of a former farm manager, Greg Rebar, "managed from the boardroom down."[33] Its properties—a hatchery on Vancouver Island and eight seapen sites in the

Broughton Archipelago—were bought in May 1992 by Stolt Sea Farm Inc., formerly Sea Farm Canada Ltd. Jacob Stolt-Nielsen Jr. had bought out his Sea Farm partner, Canada Packers, and the IBEC purchase made Stolt Sea Farm "the largest producer of Atlantic salmon in British Columbia,"[34] as well as one of the world's largest aquaculture groups. The parent company raises Atlantic salmon in Norway, New Brunswick, Maine and Washington state as well as BC; farms sturgeon in California, turbot in France and seabass in Greece; and also has fish farming interests in Spain and Chile.

One of the most complicated stories of wheeling and dealing in the entire industry involved two of the longest-running companies: General Sea Harvest, which had been bought by Cultor Ltd. of Helsinki in 1989, and Pacific Aqua Foods, which had been formed in 1985 by the amalgamation of Brad and June Hope's Tidal Rush Marine Farms, Aqua Foods Ltd. and Deluxe Seafood Ltd. In 1986 Pacific Aqua's leading shareholder, FC Financial, hired New York financial whiz Warren B. "Bud" Kirchner to revitalize the business. He brought in a new majority partner, National Sea Products of Halifax, at that time the largest fish processor in North America and producer of "Captain High Liner" products. Kirchner subsequently closed all of PAF's Sunshine Coast farms, established new ones in Campbell River, Tofino and Port Hardy, and began rearing Atlantic salmon instead of chinooks. As a result of this restructuring, Pacific Aqua turned its first profit in 1989, but a year later National Sea, which had developed financial problems, withdrew its capital. Kirchner then worked out a deal with Cultor Ltd. to manage its General Sea Harvest farms, merged General Sea Harvest with PAF, and in 1993 the renamed company, International Aqua Foods Ltd., posted a profit of $900,000.

The fish feed manufacturer Moore-Clark (Canada) Ltd. entered the salmon farming business in 1990 by purchasing Georgia Seafarms through a subsidiary, Paradise Bay Seafarms Ltd. Within five years the company owned six sites in the Quadra Island, Nodales Channel and Hoskyn Channel area. In October 1994 British Nutrition, Moore-Clark's parent, sold the company to the Norwegian fish feed manufacturer Nutreco Aquaculture, though it still continued to operate under the names Moore-Clark (Canada) Ltd. and Paradise Bay Seafarms Ltd.

Nor-Am Aquaculture Inc., headquartered in Campbell River, is one of the phoenixes of the coastal salmon farming industry. As the North American Salmon Corporation, it went into receivership in July 1989. Its financier, the Christiana Bank, sold its assets to Norent of Campbell River and Scanam of

Seattle, both of which were also financed by the Christiana Bank. After Norent went into receivership later that same year, the Christiana Bank shuffled some of Norent's assets back to a reincarnated Nor-Am. By 1995 the company was farming Atlantic salmon at sites in Frederick Arm, off Sonora Island in Young Passage and off Cyrus Rocks in Okisollo Channel. Okisollo Marketing did the company's processing at a plant north of Campbell River, and Scanam handled distribution and marketing.

The industry picture at the end of 1995 was a far cry from that of 1989. Farms were concentrated off Tofino on the west coast of Vancouver Island and off the Island's east coast from the Broughton Archipelago south to Quadra Island. The entire coastal industry was now owned by just seventeen companies, most of them subsidiaries of huge international concerns, and nine of them were supplying 75 percent of all farmed salmon produced in BC. They operated a total of 80 farms, 100 fewer than six years earlier, but these were far bigger farms than those of 1989. Automated procedures meant there were fewer employees, but staff turnover was less frequent because working and living conditions had improved.

In 1995, farmed salmon had for the first time become British Columbia's leading agri-food export, earning approximately $140 million in sales of Pacific and Atlantic salmon meat. Two years later, production by BC's farmed salmon industry exceeded that of the commercial salmon fishing industry, partly because of the farmers' increasing efficiency, though also because of the federal government's curtailment of the commercial fishery.

By 2003 BC had become the world's fourth-largest producer of farmed salmon, with twelve companies, most of them international concerns, operating 83 active salmon farms, primarily located in and around the northeast and west coast of Vancouver Island. Together these farms provided 1,800 jobs, while support industries employed another 2,340 people—90 percent of these jobs are outside the main urban areas. Sales had now risen to $310 million annually, and though initially Japan had been the major market for BC salmon, by 2003, 89 percent of production was going to the United States.

Any Fish is a Good Fish

Salmon Farming and the Department of Fisheries and Oceans

Otto Langer

Over the past 40 years, British Columbians have seen great environmental conflicts as groups and individuals stood up for the natural environment against companies and governments that wanted to exploit the province's resources. A focus of many of the battles was the iconic Pacific salmon, one of the species that define BC's past, present and future. Salmon are like the proverbial canary in the coal mine. If they survive and thrive, it indicates we have healthy ocean and river environments. When hydroelectric dams are constructed, estuaries are polluted, and loggers destroy spawning streams, salmon runs dwindle and disappear, alerting us to major problems in the whole ecosystem.

During these struggles we have learned much about salmon life cycles and habitat requirements, which allowed us to make some progress in better protecting salmon and their habitat, but still many runs and their habitats continue to decline. Habitat destruction is usually a one-way process, and when all the small impacts are added up, the salmon run dies by a thousand cuts. No one, including the various levels of government, properly addressed the protection of this common resource belonging to all Canadians. As salmon stocks declined, fishermen blamed loggers for habitat destruction. Loggers blamed fishermen for overfishing. Governments blamed each other, but all levels of government continued to allow pollution from pulp mills and sewage systems, and stream blockages by railway and road building, dyking, dam building, and mining, to mention just a few of the poorly regulated industrial impacts.

What the salmon needed least of all was yet another impact on their stocks or their habitat, but 25 years ago netpen aquaculture was imposed on the West Coast environment with the approval and encouragement of the regulators, including the federal Department of Fisheries and Oceans (DFO), which had a clear and strong mandate to protect all fish and fish habitat in Canada. The result is a resource so stressed that it seems doubtful there will still be salmon in our rivers a hundred years from now.

DFO is the government department responsible for implementing the Fisheries Act and the Oceans Act to conserve, manage and protect the fish and the fish habitat of the coastal territorial zones and the inland waters of Canada. In addition, the Fisheries Act gives it the power to control the introduction of fish and the transport and transplanting of fish. The department also has the jurisdiction to properly manage fish farming because it has to approve or issue permits for most salmon farm activities and installations—and can withhold approval or permits in order to protect fish, fish habitat and public safety on our waterways.

In 1997, parliament passed the Canadian Environmental Assessment Act (CEAA), which requires a proper environmental assessment of all projects in Canada that have an impact on federal jurisdiction or federal lands or are funded with federal money. If anyone harms fish habitat, as defined in the Fisheries Act, the CEAA process is automatically triggered by law.

With this strong mandate, what went wrong? How did DFO bureaucracy in Ottawa come to believe that salmon farms do not have any impact on the natural habitat of the West Coast? How did they decide that fish farm activity or development is not subject to the habitat protection provisions of the Fisheries Act, the Canadian Environmental Assessment Act and the Oceans Act?

Some say all organizations have their ups and downs and that this is just part of a cycle. Working in DFO, I saw my colleagues in the provincial environment department go through such cycles much more frequently, and despite my criticism of DFO as an often dysfunctional agency, it is still in much better shape than its equivalent in Victoria, which is nearly nonexistent due to legislation reversals and staff cuts by Gordon Campbell's Liberals.

However, even the anemic provincial enforcement staff has taken more fish farm transgressions to court than DFO has. DFO maintained the philosophy that any fish is a good fish, and anyone who makes more fish is a friend of government and DFO. DFO may soon try to put this into law, and one of the best pieces of environmental legislation in the world, Canada's Fisheries Act, will be undermined.

How We Got Into This Mess

Even back in 1969, when I was hired by DFO, the agency was regularly in a state of chaos. As it hired new staff to meet the environmental challenges of the 1970s, it was also downsizing to address budget shortfalls, and I soon saw there was no long-term strategic direction or funding in place. What was a priority one day could well be cut the next.

I joined DFO's Pacific Region (BC and Yukon) Special Projects Unit, which later became the first Habitat Protection Unit. The workload was clearly excessive, and many times we were forced to jettison the scientific method in order to forge ahead in the best interests of salmon and, occasionally, other species of saltwater fish. We called it good gumboots biology, and it often worked. The federal government's world-class Fisheries Research Board in Nanaimo was overly academic and did not provide us with the day-to-day scientific backup we needed for protecting fish habitat. Researchers and staff managing fish stocks did their work without taking into consideration the need to protect the natural environment. Each group did its own thing, with no connection to scientists working on separate but related questions. There was no overall planning or strategic direction, and a true team approach to problem solving rarely occurred.

Most of the habitat protection work was done in reaction to industrial development. The consequent ad hoc response seldom allowed the staff to do good scientific work or write up any data they had collected. The precautionary principle was not the order of the day; if we did not come up with the data or a good argument to support fish habitat needs, industrial destruction of that habitat was usually allowed to proceed. Lessons learned were not recorded, and the errors we made were not fully appreciated. Managers rarely looked back at what had taken place, corporate memory often failed us, and as a result, we or our successors repeated the mistakes. At the same time, many habitat protection initiatives were put on hold when finances or staff were unavailable, which set them back several years or caused them to be abandoned until someone reinvented them.

Despite this criticism, it would be unfair to say no progress has been made. Over the years I was able to work with many dedicated habitat staff, fishery officers and scientists who helped make a difference. Logs are no longer driven down our streams. Dams were never built on the Fraser River. The Greater Vancouver Regional District now treats its sewage to the secondary level before discharging it into the Fraser. The smelter in Trail no longer dumps its poisonous slag into the Columbia River. Without DFO habitat protection efforts these accomplishments would not have occurred.

However, real progress was repeatedly offset by a multitude of new impacts as BC's population grew and we cut more trees, built more roads and promoted salmon farming. Often there was only an illusion of progress that took the form of reorganization and name changes reflecting the popular language of the day. During my first three years in government, my employer metamorphosed from the Department of Fisheries to the Department of Fisheries and Forestry to a branch in the Department of Environment. Real change was slow in coming as it was difficult to develop new ways to do the job with old thinking, old tools and little or no support from Ottawa. Nor was Pacific Region DFO able to create unique approaches for the different environmental and political jurisdiction in BC. All direction comes from Ottawa. During the 1990s DFO began a series of costly bureaucratic exercises to "improve" the civil service, whether by centralizing or decentralizing staff, reviewing excessive workloads, figuring out how to do more with fewer resources, writing new job descriptions, becoming bilingual, or changing the way duties were carried out. These gave us extra work, but did little to improve things.

One reorganization that had a detrimental effect occurred in 1971, when pollution control staff were taken from DFO and moved to the new Department of Environment (DOE). This meant that people responsible for fish habitat protection were no longer responsible for the quality of the water that the fish lived in. DFO still depends upon DOE to protect water quality for fish, a jurisdictional split that has undermined fish protection for over 30 years.

This might not have been a problem if DFO and DOE had been able to work together, but the two departments frequently disagreed on what was a problem. In particular, DOE did not, and still does not, have a feeling of responsibility for the fisheries resource, even though it was given a key part of the Fisheries Act to administer. There was constant buck-passing between the two agencies as well. In 1980 I was responsible for bringing DOE, DFO and BC Ministry of Environment staff to a site on the lower Fraser River where a cement ready-mix facility was dumping waste concrete onto an estuarine marsh. The concrete mix and wash water discharge was highly toxic to fish, and the concrete filled in valuable fish habitat. However, the DOE representative maintained it was a habitat (DFO) problem. DFO insisted it was a pollution problem that DOE and the BC Waste Management Branch (WMB) should address. The BC conservation officer noted that it was not a pollution problem—the company was simply reinforcing the dyke! The provincial staff then told the company that it should apply for a pollution permit that could allow it to dump concrete legally.

Thirty-two years after the DOE and DFO split, the Fisheries Act buck-passing was still taking place. In 2003 I sent copies of the same letter to the ministers of DFO and DOE. It outlined the lack of action on salmon farmers' illegal use of chlorine to disinfect their farms after a series of disease outbreaks. DOE minister David Anderson noted that such issues were a DFO responsibility. A week later DFO minister Robert Thibault responded, noting it was a DOE responsibility. No one was willing to take a lead in addressing the discharge of this highly toxic chemical into water frequented by wild fish. Despite the fact that many believed David Anderson was strong on environmental issues, his record addressing salmon farm problems was dismal as a DFO and then the DOE minister.

As the cement facility example indicates, there were also conflicts between federal and provincial people, and this had a major effect on DFO's approach to fish farms. The economic drive to develop BC and create more wealth

Aerial photograph of an ocean net cage in the Broughton Archipelago.

always took precedence over habitat protection. The WMB seemed to believe that its mandate was to issue a permit to pollute the environment to anybody who wanted one. Its stated position was that the assimilative capacity of a waterway was a renewable resource, and it should be used to reduce costs to industry. In other words, lowered water quality was to be traded for more jobs.

Federal–provincial relations were always high on the agenda, and issues like proper environmental management were regularly subordinated to the priority of keeping the province happy. This was most unfortunate in that the Canadian constitution makes federal legislation, such as the Fisheries Act, superior to provincial legislation. Subordinating national interests to the provincial political needs of the day does not serve the long-term interest of Canadians.

The driving imperative behind most provincial and federal decisions and programs was the need for greater economic activity and job creation. BC governments of all stripes lobbied aggressively to ensure that federal staff and the needs of fish did not hamper their economic agenda. Protection of fish and fish habitat became a more distant priority for DFO, even though in 1986 the federal government adopted a national habitat policy to ensure a "net gain" of fish habitat. Sadly, the environmental enforcement record goes up and down like the tide in Prince Rupert. For most of the past 30 years, DFO and DOE have failed to protect habitat or water quality when it would conflict with the needs of the provincial government or industry, and both levels of governments are now working to formally establish a model of industry self-policing to address the downsizing they have forced on environmental agencies. Unfortunately, this approach has never worked well, even when those involved have the best intentions. The drive to make money will get in the way of environmental stewardship in most businesses.

Money was also an issue for DFO. Between 1992 and 2003 the department received tens of millions of dollars for special projects, such as the Fraser River Green Plan, the Habitat Action Plan, and the Canadian Fisheries Adjustment and Restructuring Plan, but in order to maintain old programs and staff, DFO constantly took money from new programs to support old programs, thereby making the new initiatives less effective. By the late 1990s the habitat management staff had increased in size; fishery biologists were responsible for implementing new legislation, such as the Canadian Environmental Assessment Act, the Oceans Act, the Species at Risk Act and the Navigable Waters Protection Act (during yet another reorganization in 2004, responsibility for NWPA was assigned to Transport Canada, though

DFO will still have to review each application under the act); and the Canadian Coast Guard also became a DFO responsibility. At the same time, from 1998 to 2004, DFO had to cut large numbers of staff to help eliminate the federal deficit. The DFO research branch lost about 40 percent of its staff, and too often it was the good staff who took buyouts. The deadwood stayed behind to work on their pensions. The department was now doing a 21st-century job with the same budget it had fifteen or twenty years earlier.

Finally, DFO was caught in a quagmire of poor leadership at the political and the senior bureaucratic levels. Ministers were constantly changing as cabinets were shuffled or governments changed, and they often seemed desperate to get out of the Fisheries portfolio. Managers were shifting as well, promoted because of their connections in Ottawa and not because they had any training in, or a complete understanding of, their job. A hovercraft navigator was promoted through the system to be the director of the lower Fraser River region—scene of BC's most controversial fishery conflicts. Another Coast Guard officer from Ottawa was put in charge of managing the fisheries resource in the Skeena River. A salmon biologist was promoted through the system to become the director of the Canadian Coast Guard. To become a director you had to spend a year or two in Ottawa, where you were taught that Ottawa always knew best and federal-provincial or federal-industrial relationships were more important than the natural resources DFO was established to manage—the fish and their habitat.

Staff were dizzy from never-ending reorganization. Their mandate became ever more complex as resources shrank, and in the midst of no clear direction, the salmon farming industry was heating up to expand.

DFO and Fish Farms in the 1990s

During the 1990s the salmon farm industry was subjected to greater scrutiny than ever before by the public, better-organized environmental organizations, fishermen, First Nations and the media. This scrutiny was well deserved. The mom-and-pop operations along the coast had disappeared, and multinationals had taken over the industry. The problems documented over the previous years (including ocean bottom destruction caused by waste food and feces; disease; chemical pollution of the marine environment; navigation hazards; and the escape of Atlantic salmon) had become full blown, but the industry was expanding production, and government and industry were not addressing the known problems. Denial was the order of the day.

In part this was because the industry was now dominated by big investment and big business. To make salmon farming a profitable venture, these business people and investors had successfully lobbied the BC and federal governments to allow them to introduce a "domesticated" farm fish—the Atlantic salmon—to the West Coast. Dr. David Narver, director of the BC Ministry of Environment, Lands and Parks (MELP) fisheries branch at the time, and DFO director-general Pat Chamut were opposed to introducing foreign Atlantics to the West Coast. Over 200 years of lessons learned when alien species were introduced to natural habitats should have warned government that this was a dangerous conservation move. However, the higher political levels insisted that the farm industry could have its Atlantic salmon and it was the job of DFO and MELP staff to ensure that it could be done safely. Politics, rather than science, determined the posture for the management of salmon farming on the West Coast.

In 1995 the federal government decided it was a national priority to promote aquaculture in order to support economic development and replace the fish lost by mismanagement in Canada and elsewhere in the world. Without passing any empowering legislation, the government dumped this new responsibility on DFO, which created an independent office of aquaculture development in Ottawa and gave it a generous budget to promote fish farming. This office, which was basically an industrial lobby group supported by the taxpayer and reporting directly to the Minister of Fisheries, frequently undermined the conservation intent of the Fisheries Act.

In 2002 DFO's chief of aquaculture in Vancouver, Ron Ginetz, was pulled out of the fisheries branch and transferred to the habitat and enhancement branch. This did not sit well with him as his role was to promote aquaculture and he always battled with the habitat staff over their concern that habitat was being harmed by fish farming. To the astonishment of many DFO staff, he then succeeded in getting a secondment to the BC Salmon Farmers Association (BCSFA), where he continued to work as a federal civil servant until April 2004, receiving a cheque from the federal government every two weeks and the full benefits of a federal civil servant, even though his new job was often at odds with DFO's mandate.

Fish farm wastes

Did DFO not know what salmon farming and other forms of fish farming could do to fish habitat, wild stocks and the ecosystem? In a DFO habitat protection meeting during 1990, two experienced biologists, Rob Russell and Jim Morrison, showed staff the impacts of open net cage salmon farms. Both men were divers, and their underwater photos showed that fish feces and uneaten fish food had accumulated to a depth of over 30 centimetres, smothering all marine life under the fish farms they had inspected. However, enforcement officers were never called in, and the industry was allowed to continue dumping its waste, just as the ancient Ocean Falls, Port Alice and Woodfibre pulp mills were allowed to fill in the ocean bottom with lost woodfibre 50 years earlier.

Staff were confused. They had a Fisheries Act and a "no net loss" policy that clearly did not allow this habitat destruction, but due to constant DFO reorganization, the enforcement staff had higher priorities, such as working on the salmon and herring fisheries and aboriginal issues. In addition, DFO headquarters was promoting fish farming. Any time anyone could make a fish, it was a good fish. Nothing was done to change the status quo, and this new industry was allowed to continue polluting, like industries that started up many decades earlier when there were no environmental controls. Salmon farms seemed to have been given immunity from the Fisheries Act.

The public was soon well aware of the problem of fish farm wastes, and in 1999, Lynn Hunter of the David Suzuki Foundation laid Fisheries Act charges, as a private informant, against a Stolt salmon farm that had smothered fish life on the ocean bottom. In such cases the provincial or federal Attorney General's office can intervene and take over the action or derail it. BC passed the case to the federal Department of Justice (DOJ), which took the position that an obvious violation had occurred. However, due to the inept handling of the matter by DFO there was a reduced likelihood of a successful prosecution. DOJ legally stayed the matter.

The problem was that DFO had been working with the province on fish farm pollution rather than dealing directly with the company. The company had a valid licence from the province to carry out its business and could legally claim that it did everything that was allowed in its provincial permit and was not aware of different Fisheries Act or DFO requirements. This could give Stolt a strong defence on the grounds of due diligence or officially induced error.

For years DOJ had urged DFO to directly notify those harming habitat that they were contravening the requirements of the Fisheries Act. DFO could also have required that Fisheries Act habitat protection needs be specified in the provincial permits. DFO had done absolutely nothing about it, even though the farm was destroying habitat, and this gave the company a defence from prosecution under the Fisheries Act. DFO put the political need to work with the province ahead of responsible habitat protection. Despite such a public embarrassment, DFO did not begin to enforce its legislation or carry out its legal responsibilities until 2003, when enormous pressure from the public and even DFO staff forced it to begin to relate to its own legislation. However early indications are that little has changed in terms of habitat protection.

Atlantic salmon and seals

Since the introduction of Atlantic salmon twenty years ago, DFO has regularly stated that Atlantics are no risk to wild Pacific salmon. In the late 1990s environmentalists initiated an access-to-information search that uncovered an internal memorandum, dated February 1991, in which DFO chief of aquaculture, Ron Ginetz, wrote to the fisheries branch director, Al Lill: "In my view it is only a matter of time before we discover that Atlantics are gaining a foothold in BC (residency) or in Washington state. Even if agreement cannot be reached on this assumption, what should our position be in responding to the enquiries? Do we prepare public/user groups for the possibility, and strategically plant the seed now, or do we downplay the idea and deal with the situation if and when it occurs?"

Seals looking for an easy meal are targets of fish farm staff.

The memo also raised the issue of "seal dispatchment." Ginetz noted that, as expected, fish farm employees were killing more seals and sea lions, which hung around the farms' netpens, snatching fish for an easy snack. DFO issued five-dollar permits that allowed fish farmers to kill as many seals as they wanted, and the agency would need to develop a strategy to deal with the media on this subject. The DFO predator control permits allowed salmon farmers to kill seals and even Steller's sea lions (an endangered species) when they attacked the farm nets to get at the big fat salmon swimming inside. The farmers were required to report all kills, dispose of the corpse properly and submit samples, but often they did not report numbers killed, did not submit samples, disposed of killed animals improperly, killed seals that were well removed from their farms, and did not comply with Canadian firearms control laws.

PHOTOGRAPH BY ED MAY

An illegal dump of seals killed by salmon farm employees was found in Clayoquot Sound in the spring of 2001.

When environmentalists submitted a follow-up request for information, DFO was unable to find Lill's response, which did not surprise me. When you read between the lines of bureaucratese, you can translate Ginetz's memo into an obvious decision in plain English. Ginetz knew that DFO would not warn the public that its decision to allow the introduction of Atlantics to this coast could be a risk to wild salmon. Also, it would not admit that it had made a mistake. The memo was designed so Lill could endorse the approach of keeping everyone in the dark, hope like hell it did not get out of control, and spin any issue that did hit the media.

The question about the seal kill was similar. It did not indicate that DFO's five-dollar permit was unacceptable and must be changed. It just said that this was an issue that could embarrass DFO, so they had to control the information that got out so they could avoid looking bad in the media.

Ten years after this memo was written, DFO was still downplaying the threat of large escapes of Atlantic salmon, and the agency had to be taken to court to get it to admit that its predator control permits were not being enforced and had allowed the relatively uncontrolled killing of seals and sea lions.

Through a combination of such undermining of the public interest, lack of vision, inept leadership, continual reorganization and disorganization, political promotion of the industry, and now near insolvency in DFO, we arrived at the current fish farm mess. Recent DFO actions to address its self-created embarrassments do little to address the problem. In 2003, DFO's response to demands for quicker approvals of fish farm licences was another reorganization. There is now an aquaculture director whose job is to develop good working relationships with the province, promote the industry, and direct the habitat protection and environmental assessment staff for all fish farms. DFO also brought in a staff member from the provincial Ministry of Agriculture, Food and Fisheries (MAFF) and hired a fish farm worker to help in its review of issues related to salmon farms. This reorganization produced a complete conflict of interest. Those who assess the impacts of farms or enforce violations should work at arm's length from those who facilitate and expedite the industry. DFO does not seem to understand that it cannot promote salmon farming and regulate it at the same time.

The Provincial Approach to Fish Farms

When the New Democratic Party came to power in BC in 1991, it saw fish farms as a potential job generator and moved the old commercial fisheries branch out of the environment ministry and into the agriculture ministry so it could promote fish farms. However, by the mid-1990s the failures in fish farming and the public outcry against the industry forced the government to recognize that various problems had to be addressed before the industry was given more encouragement to expand.

Urged by staff in the provincial Ministry of Environment, Lands and Parks, the NDP initiated the Salmon Aquaculture Review (SAR), which released a comprehensive report in 1997. The SAR scientific panel looked at all aspects of salmon farming and made numerous recommendations to address the industry's problems, which could have been implemented during the five-year moratorium that followed. Unfortunately, there were significant loopholes in the moratorium, and salmon production doubled on existing farms while it was in place.

In 2001 BC elected a Liberal government, and within a year the Liberals announced that they were lifting the moratorium. However, this was delayed because DFO and MAFF couldn't agree on a common approach to fish farm issues—especially the protection of fish habitat from the effects of open net cages. For instance, the province, in cooperation with industry, had established what it believed was an acceptable level of farmed fish feces that could be deposited on the ocean bottom by measuring the quantity of poisonous hydrogen sulphide that the waste generated. Because this approach would allow significant destruction of habitat and marine life, DFO could not agree with this made-in-BC standard.

BC forced the issue and unilaterally moved to encourage development of new fish farms as part of a coastal development strategy. The provincial government did not accept environmental impacts as a major concern. Al Martin, a past director of aquaculture in MAFF, stated that the BC position was to promote the industry and manage the problems as they arose. In 2004 this official position was refined by the assistant deputy minister, Bud Graham, who said: "The risks caused by salmon farming are not that great and are clearly manageable." This is despite the fact that more information is now available on the negative environmental impacts of salmon feedlots in public waters.

Once again DFO, supposedly the steward of the public interest and the legal protector of fish and fish habitat, abdicated its leadership role and did not resist political and industry pressure. It went along with the BC government's lifting of the moratorium, even though there was no agreement with the province on how the industry would be managed and what standards would be in place. DFO had previously agreed that the province would regulate what occurred on the fish farms and DFO would protect the environment outside the farm. This meant BC was allowed to assume the mandate to protect the habitat directly associated with fish farming. DFO would protect wild fish, naively believing you could separate wild fish from fish farms, even though they lived in the same habitat. It is impossible to protect fish and their habitat when you give up the ability to control what goes on in a fish farm, but once again a partnership to keep a province happy was most important to Ottawa, and once again management agreed to ignore DFO's mandate, or water it down, and live with BC's requirements.

In 2000, in the wake of Lynn Hunter's private informant charges against the Stolt farm, Bruce MacDonald, a competent DFO habitat chief in Nanaimo, decided that DFO had to start with a clean slate. The embarrassment and inaction had to be reversed. He issued a legal authorization agreement under the harmful alteration, disruption or destruction provisions of Section 35 of the Fisheries Act to a Stolt salmon farm. When a company cannot avoid harming habitat as it develops its business, it can apply to the DFO minister, who can legally authorize damage to habitat subject to conditions such as "no net loss" and monitoring. After much internal debate in DFO, the authorization was ordered withdrawn. With DOJ input, DFO drafted a new authorization and forwarded it to Stolt for signing. The BC Salmon Farmers Association called up Yves Bastien, DFO's commissioner of the office of aquaculture development, who got in touch with his old friend Guy Beaupre, director of the habitat and enhancement branch (HEB) for the Pacific Region. Beaupre, acting with his aquaculture chief, Ron Ginetz, ordered staff to again recall the authorization. He and the bureaucrats in Ottawa determined that DFO could not prove that a salmon farm would harm habitat, despite the clear and abundant information collected over the previous decade that indicated open-net fish farms could have significant environmental impacts.

Guy Beaupre exemplifies the problem I mentioned earlier of managers being promoted because of their connections in Ottawa and not because they had any training in or understanding of their job. He was an economist who had worked in the Department of Finance and the Prime Minister's Office before being transferred to Ottawa DFO to work on federal–provincial relations. In late 2001 Ottawa decided to send him to the Pacific Region to direct HEB. Other than his understanding of the need to keep the industry and the province happy, he had no knowledge of the West Coast, the fishery or habitat. He was also the eighth acting director of HEB in six years. It would seem that habitat protection was not a priority for DFO, as an important position would not have been treated as though it were a game of musical chairs. In fact, the branch had been undergoing a constant reorganization over that six-year period.

After a multi-day meeting in Victoria with the BC MAFF staff and the BC fish farmers in December 2001, Beaupre agreed that DFO would expedite all applications for fish farm approvals on file within six weeks and spelled out the advantages of collaboration and consultation with the province and industry. As part of his initiative to work better with the province, he seconded an aquaculture worker from MAFF to act as the new DFO chief of aquaculture. Just as royal families intermarried to build alliances, DFO allied itself with MAFF and the fish farmers by bringing a fish farm promoter into the fold. It was to cement these alliances that Ron Ginetz was sent to work for the salmon farmers.

In early 2002, Guy Beaupre called a meeting of senior DFO staff to discuss the habitat protection, salmonid enhancement and aquaculture issues the agency was dealing with at that time. He described the productive meeting he had just had with MAFF and the salmon farm industry in Victoria. I was still working for DFO at the time, and I tried to impress upon Beaupre that there was a public resource and interest at stake, and that meetings should include representatives of the fishing industry and the environmental community to speak on behalf of the wild fish and the ecosystem. I asked him who was protecting the public interest—and I am still waiting for an answer.

Near the end of the meeting, Beaupre warned staff that CBC TV's *Nature of Things* program was going to run a show on salmon farming. He noted that the show, hosted by David Suzuki, was full of misinformation. I asked him if he had seen the show. He said that he had not, but Anne McMullan, executive director of the BCSFA, had told him about it. I said that David Suzuki was an accomplished and internationally recognized scientist, and I asked Beaupre what his or Ms. McMullan's science credentials were. His reply was that Anne McMullan was a very nice person, trustworthy and someone DFO could work with.

For me, this was a final confirmation that DFO had lost sight of the issues and its mandate, and I knew that the situation in the agency would get a lot worse before it started improving. Bureaucracies frequently function that way. I was in my mid-50s and no longer had the patience to put up with inept leadership, endless and mindless reorganization, and vanishing resources in the hopes that a better day would arrive. I could do my job more effectively elsewhere. A few months later I left DFO and went to work for the David Suzuki Foundation as director of a new program on marine conservation.

Then things got worse.

Sea Lice and Dead Fish

In June 2001, Alexandra Morton, a whale researcher and writer, discovered that sea lice in the Broughton Archipelago were infecting and killing juvenile pink salmon that had just emerged from their natal streams and had to pass dense concentrations of salmon farms to reach the ocean. DFO did little to respond to this alarm other than deny it. Six weeks later it sent a charter boat out to the Broughton Archipelago and sampled fish—using a net that was the wrong size and would dislodge most sea lice attached to the pink salmon. The researchers looked for the fish at the wrong time, the wrong depth, missed the main area of complaint and then noted there did not appear to be any sea lice problem.

This shoddy work, and continued public statements denying fish farm impacts, further polarized environmentalists and others concerned about fishery resources, including First Nations. They realized they had to get organized to address a problem that appeared to be getting worse every day. A large contingent of environmental groups including the David Suzuki Foundation, Living Oceans, Watershed Watch, Rain Coast and aboriginal groups formed the Coastal Alliance for Aquaculture Reform (CAAR). They aimed their anger at DFO because they saw the province as a fish farm promoter without the legal authority to do what it was doing. DFO, on the other hand, had the authority to regulate or prevent the negative impacts of fish farms, but was not acting.

In December 2002, representatives of CAAR met with John Davis, regional director of DFO in BC. At that meeting Davis argued that everyone had to understand the situation DFO was in. The environmentalists wanted better monitoring of the industry, and the industry wanted DFO to expedite its expansion plans. Davis noted that it was DFO's job to find a balance between those two positions. He clearly saw his role as political compromiser and did not appreciate that he was legally required to implement and enforce the strong Fisheries Act legislation that was in place.

This response was not unexpected. Several years earlier, when he was DFO's director of science, Davis intervened in a dispute with the David Suzuki Foundation over a salmon farm issue and allowed BCSFA to reprint his letter as a large paid ad in the *Vancouver Sun*. The DFO bias was obvious. As a DFO employee at the time, I was shocked that DFO would allow itself to be used in such a manner by the industry it had to regulate. Davis also directed Dr. Don Noakes, who regularly took the lead to protect the industry from criticism and had become DFO's apologist for the industry. He was also the DFO scientist in charge of aquaculture research. Only DFO could explain why it allowed such a blatant conflict of interest when the public expected it to provide strong leadership.

In February 2001, DFO and MAFF allowed workers at the Heritage Cliff Bay farm in the Broughton to load two large seine boats with farmed salmon infected with IHN—a lethal viral disease. These fish were then transported into the Fraser River estuary to be unloaded for rendering. There were no precautions in place at the rendering plant to prevent infected blood, scales, slime

PHOTOGRAPH BY ALEXANDRA MORTON

Morts, or dead farmed fish, are removed from a Sir Edmund Bay farm site in the winter of 2003, during an IHN outbreak.

and body parts to run back into the Fraser River—the world's largest salmon river. Despite many pleas by the David Suzuki Foundation to stop this reckless activity, neither DFO nor MAFF would do anything about the problem.

At nine that night, as the boats filled with diseased fish were coming into the river, the David Suzuki Foundation and the Sierra Legal Defence Fund found a judge who granted an injunction to stop the fish from being unloaded. DFO fishery officers on night shift refused to deliver the injunction order. They said senior management had directed them to ignore the injunction. The RCMP said they would not deliver the injunction to the boats waiting to unload because that was the job of DFO fishery officers. Some lower-level DFO staff were shocked that their agency would have allowed this to happen.

A few days later, Heritage lawyers had the injunction set aside by agreeing to an undertaking that the company would not allow any contamination of the Fraser River. As the fish rotted in the holds of the boats, Heritage decided to play it safe, transporting the fish to Vancouver Island for composting. The company seemed to learn more quickly than the regulatory agencies. In this matter, MAFF and DFO were clearly negligent in their job of protecting fish stocks and the public interest. Their judgment was impaired by their political need to adhere to the agenda of promoting fish farms at almost any cost.

It is ironic that DFO and MAFF have guidelines in place to prevent the contamination of other fish farms and even fish at processing facilities, but they totally ignore the possibility that disease may spread to wild fish stocks. Bureaucrats stated that IHN was found naturally in the wild, and they saw no problem with its being introduced into BC's largest salmon river. In fact, outbreaks of IHN in the wild are very uncommon.

The incident had little impact on DFO. The next fall, with DOE, it participated in issuing a permit to a Vancouver Island salmon farm to dump over 1,000 tonnes of dead and rotting salmon into the Pacific Ocean off Nootka Sound. The dead fish contaminated an area rich in bottom fish, and a week later they were drifting onto the western Vancouver Island shoreline. The groundfish fishermen were livid that rotten salmon would be legally dumped onto a designated fishing area.

In 2002, DFO's Ottawa and regional managers inexplicably gave in to Department of Justice and staff pressure and allowed habitat protection staff to use the habitat provisions of the Fisheries Act and the Canadian Environmental Assessment Act to assess applications for new fish farms. This was only sixteen years after the "no net loss of habitat" policy was adopted and 28 years after the habitat protection provisions were proclaimed as part of the Fisheries Act. In 2003, DFO completed its first full CEAA review of a new fish farm in the area of the Broughton Archipelago plagued by sea lice. DFO had finally acknowledged that salmon farms could harm habitat and should be subject to environmental review, but the staff still decided that it would be all right to permit the establishment of a new fish farm at Humphrey Rock. Allowing another farm in an already stressed area demonstrates a questionable understanding of ecosystem functioning and cumulative effects.

The review for the Humphrey Rock farm was completed just as the adult pinks were returning, eighteen months after Alexandra Morton had raised the alarm about sea lice on juvenile pinks. The run came back at historically low levels. A DFO population expert noted that such a collapse of salmon had never before been seen on the West Coast, and the epicentre of the collapse was in the area of the salmon farms. He was not heard from again.

In response to the embarrassment of this catastrophic wild pink collapse, DFO spent over a million dollars to monitor the situation, but in January 2004, scientists noted that their work was not designed to examine the link between sea lice, wild salmon and fish farms. In early May 2004, they sent out a press release noting that they saw no relationship between salmon farms and sea lice, or between the presence of sea lice and the loss of young salmon. It's not clear if the narrow focus of pure science caused this obvious part of their work to be omitted or if there was a political agenda directing the DFO research. But it is clear that the public expects more objective and relevant work from an agency that was a world leader in fishery science before it was captured by industry and politicized to support the government's aims.

I could relate dozens of other stories showing that DFO lacks the staff or the will or the support in Ottawa to do the job taxpayers are paying it to do. However, further examples would only confirm the message that, starting about twenty years ago, DFO has failed to regulate salmon farming in a responsible manner. It has recently been embarrassed into applying bandages for serious internal conditions, but something more like major surgery is required to breathe life into an ethically moribund agency.

In 2004, CAAR met with Jean Claude Bouchard, an assistant deputy minister of DFO. He had been meeting with MAFF staff and set aside half an hour to meet with CAAR, which was raising hell over DFO's pathetic record on fish farm management. Bouchard made it clear that DFO would not do anything drastic to change things. He noted that even if he were convinced that open net cages should be phased out, no one in Ottawa would agree with him.

CARTOON BY ADRIAN RAESIDE

When Paul Martin became prime minister in December 2003, he named a new DFO minister, Geoff Regan, a lawyer with no knowledge of or experience with the fishery. I expect that he will quickly learn how to maintain the status quo. A March 2004 press release confirms that he will believe what his higher level bureaucrats and the office of aquaculture development tells him—support aquaculture and remove any impediments that hold up development of the industry. Little will change.

Blueprint for Change

If there is to be hope for the wild fishery of Canada, fairly simple changes are needed, but they will not be welcomed by an entrenched bureaucracy with an obviously politicized aquaculture agenda. We need strong and enlightened leadership in the minister's office, in tune with DFO's mandate, the requirements of the Fisheries Act and the intent of the Canadian Environmental Assessment Act. We need staff with a will to sustain and conserve our biological legacy and to carry out the responsibilities Parliament has given DFO.

DFO cannot do its job until the government begins to hire or promote managers based on expertise, experience and merit. Currently DFO has about 30 directors in the Pacific Region. At least half of them do not have knowledge or experience in the areas they manage. The federal government believes that any manager can make a decision after a good 20-minute briefing, which is why it will parachute in directors with no fishery expertise. There will be no fisheries recovery and no improvement in the regulation of salmon farming until there is a management recovery program for DFO. Simply put, to rehabilitate the fishery resource we must first rehabilitate DFO.

Managers must understand the past to plan for the future, but DFO no longer has any corporate memory. New managers in DFO frequently refuse to depend on experienced staff for advice, insisting that they are planning for the future and do not want to revisit the past. This means they have no access to the information base and lessons learned in previous years. Instead, these managers rely on direction from Ottawa, and the fish, habitat and public interest suffer.

The federal government and the minister must stop strangling DFO with significant resource cuts. In 2004, about 40 percent of the habitat staff was cut. Those remaining will be expected to do double the work in a job that is four times as complex as it was ten years ago. And they will be doing it with the resources they had twenty years ago. An agency that is in continual contraction and in a constant flux of reorganization and disorganization will continue to be plagued by many of the problems discussed in this essay. Most Canadians expect better. This includes everyone from the salmon farmers to those who are opposed to many of the practices of the salmon farmers.

Science must be clearly separated from the political needs of the day. In the 32 years I was in DFO, the line between a scientific conclusion and a conclusion needed to make a certain government agenda look good has grown more vague. Things are now so confused that many people, including those at the technician level, interpret what the minister wants as predetermining what they can or can't do or say. This has undermined the intent of the Fisheries Act and the integrity of the aquatic environment. A few decades ago the Fisheries Research Board of Canada was separate from politics. That board was dissolved and now DFO research scientists report to the bureaucracy and are subject to the direct and indirect political needs of the day. We need more diligent scientists who will stand behind their data and refuse to be used to support the government's political needs. We must establish a new civil service culture and ethic, which shows scientists and others that they can get ahead through technical and managerial merit and not by sucking up to the senior bureaucrats and politicians or by stepping down on junior staff.

DFO has a large troop of uniformed fishery officers who can do good work if they are motivated and given the opportunity and resources needed to do their job. However, these are rarely available, and as a consequence, DFO's record of habitat enforcement is hit and miss. Industries supported by the government, like aquaculture, seem to be immune to enforcement. Salmon farming is just one of many areas in which DFO has neglected its responsibilities. In spite of having powerful legislation, DFO has failed to protect streams from logging over the past 50 years, and farming, mining, and other industries have added to the cumulative stress on the fishery resource. Adding fish farming and possibly oil and gas exploration to the BC coast will only increase the dilemma the resource faces.

Canada has signed on to international protocols, such as the Rio and Johannesburg agreements, supporting biodiversity, ecosystems and the adoption of the precautionary approach. These agreements commit Canada to take a more holistic and risk-averse approach to managing our fishery resources and environment. However, despite the passage of recent legislation such as the Canadian Environmental Assessment Act, the Oceans Act and the Species at Risk Act, the importance of a holistic environmental protection approach to insure a healthy and intact ecosystem for future generations is not recognized because of the myopic approach driven by narrow, short-term thinking.

The government will do the least it has to do to convince the public and the international community that it is proactive in these areas. The resulting minimalist approach produces narrow pieces of legislation and actions that dissect the ecosystem into separate parts, and leads to a lack of commitment to and understanding of international commitments and principles that must give rise to the protection of entire ecosystems. This is most unfortunate for a nation that should lead the world in environmental sustainability.

If Canadians see the fishery and fish habitat, the aquatic environment, and even clean water as an important part of their life, many of the impediments to maintaining and restoring that key part of our environment must be removed. Should this not happen, the aquatic environment, with its great renewable natural wealth, will continue to be degraded and wild salmon may not be part of our children's future heritage. Unfortunately, the economic agenda may be more attractive in a business-driven society governed by politics with a short-term agenda. Government will continue to do what is necessary to appeal to industry and support environmentally threatening industrial development and greater consumerism, while mouthing platitudes about environmental protection and sustainability. It will be impossible to repay the environmental debt that is passed on to future generations.

Silent Spring of the Sea

Don Staniford

Rachel Carson's classic book *Silent Spring* (1962) first raised public awareness of the environmental risks of human-made chemicals such as DDT, PCBs and dioxins. In her less well-known previous books, *The Sea Around Us* (1951) and *The Edge of the Sea* (1955), she marvelled at the wonders of the ocean.[1] A half-century later the sea has become a sink for a cocktail of cancer-causing chemicals and contaminants. Rivers carry pollutants from agricultural runoff, industrial discharges, hazardous wastes and human sewage. Pesticides used in terrestrial farming systems are so noxious that they have wiped out whole river systems. Hormone-disrupting compounds can even change the sex of fish. The sea, some seven-tenths of the earth's surface,[2] is the world's waste depository—our toxic toilet.

We have polluted our marine environment to such an extent that we are literally reaping the consequences via the bioaccumulation of contaminants up the food chain. In the Northern Hemisphere hotspots of chemical contamination—the Baltic, Mediterranean and North seas—wild fish from European waters are now eight times more contaminated than those from the South Pacific.[3]

As well, the World Health Organization has issued a caution acknowledging elevated pollution levels in farmed fish, saying "the risk of consuming contaminated fish must be weighed in view of the beneficial nutritive effects of fish."[4] A paper published in January 2004 in the prestigious scientific journal *Science* revealed that farmed salmon was contaminated with fourteen cancer-causing chemicals including DDT, PCBs, dieldrin, dioxins, chlordane, toxaphene, lindane and hexachlorobenzene. Researchers found that farmed

salmon from Scotland, the Faroe Islands and Norway were so contaminated that it is safe to eat only three to six servings a year.[5] The salmon farming industry likes to portray itself as an innocent bystander caught up in the crossfire, but it knew as early as the 1970s that the fuel supply for farmed salmon (fish oil and meal from wild-caught fish) was contaminated with carcinogenic chemicals—well before the latest revelations in *Science.*[6]

However, the contamination issue represents only the tip of the iceberg. Not all the chemicals deposited in the sea are there by accident. Some of the same chemical companies that Carson described in *Silent Spring* as mounting a "crusade to create a chemically sterile, insect-free world" now peddle their wares for use in the sea in aquaculture.

The list of chemicals deployed since *Silent Spring* was published reads like a litany of crimes against nature: canthaxanthin, dichlorvos, azamethiphos, cypermethrin, teflubenzuron, ivermectin, emamectin benzoate, TBT and malachite green, to name those put under the microscope here. The chemical conveyor belt has been well stocked by some of the world's largest chemical companies: Novartis (Ciba Geigy), Hoffmann-La Roche, Bayer, Unilever, Merck Sharp Dohme, Norsk Hydro, BP, Shell, American Home Products, Cynamid and Schering Plough. As Carson so presciently warned back in 1962: "What we have to face is not an occasional dose of poison which has accidentally got into some article of food, but a persistent and continuous poisoning of the whole human environment."

Welcome to the "Silent Spring of the Sea."

The Chemical Arms Race

Many of the pesticides, insecticides and fungicides used by salmon farmers are derived from the Second World War's chemical weapons programs[7] and the agricultural sector. As salmon farming expanded rapidly in the 1970s and 1980s, so did its appetite for new chemicals.

Just as intensive agriculture uses chemicals to treat diseases and parasites, so too does the aquaculture industry. The crucial difference between agriculture and aquaculture is that some common chemicals used in sea cage salmon farms were intended for use on land, not in the sea. Sea lice breed in the billions on factory salmon farms and are to salmon what ticks are to cattle and sheep. Yet chemicals designed for use on terrestrial livestock such as chickens, sheep and cattle simply are not suitable for use on aquatic species such as salmon. Even on land these chemicals are highly toxic, but in the marine environment their effects are magnified.

Shellfish in particular are considered collateral damage in salmon farming's "War on Sea Lice."[8] Chemicals that effectively kill sea lice (until the lice build up a resistance) also affect other members of the crustacean family—including lobsters, crabs and shrimps—and other aquatic species like oysters and mussels. Besides causing immediate death, the chemicals can also produce paralysis, premature moulting and impotence in shellfish. "It is all very well to say that the fish farmers want to hit the sea lice before they spawn in March. But this is when shellfish such as crabs, lobsters and prawns also spawn, and the treatments used could hit them as well," says Hugh Allen of the Mallaig and North West Fishermen's Association in Scotland.

Salmon farmers' response to the disease and parasite problem has typically not been to reduce stocking densities, scale back on production or leave the sea bed fallow so it can recover. Instead they have resorted to an ever more powerful arsenal of dangerous and hazardous weapons, pressuring agricultural chemical companies to develop novel treatments for use in aquaculture. If chemicals do not exist or are not suitable for use in the sea, some salmon farmers merely use what is available, even if it means breaking the law and ignoring manufacturers' labels indicating the chemicals are marine pollutants, not to be used near water, let alone in water.

The illegal use of chemicals is only part of the problem. Governments' role in legalizing toxic chemicals for use in the sea is tantamount to state-sponsored pollution and borders on corruption. Chemical companies have lobbied so successfully that the decision to license chemicals is based more on political and economic expediency than consumer or environmental safety. Often governments have granted licences for toxic chemicals to be used on salmon farms before proper scientific risk assessments are carried out, or after assessments based not on rigorous (and expensive) field trials but on simulated modelling—sacrificing science for speed. There is a time lag between approval of chemical use and the publication of peer-reviewed environmental risk assessments. Scientific papers on the environmental impacts of the carcinogenic organophosphate chemical dichlorvos, for example, started to surface in the 1990s—some twenty years after it was first used, and in some cases after it had stopped being used altogether.[9] Chemicals such as cypermethrin and emamectin benzoate have been in use since the 1990s, but risk assessments are only now being published in peer-reviewed journals.[10]

Other environmental risk assessments are deemed so controversial that they remain under lock and key, marked "Private and Confidential" and protected by client confidentiality clauses. For example, despite beginning ecotoxicology work on teflubenzuron in 1995, Nutreco (owners of Marine Harvest—the largest salmon company in the world) had, at time of writing, not published any peer-reviewed environmental risk assessments.[11] Other documents are coded so the name of the chemical is unknown.[12]

Where the name of the chemical is given, it is sometimes impossible to penetrate the veil of secrecy. For example, the original documents submitted by Ciba Geigy in the 1980s to secure a licence to use dichlorvos are still out of bounds.[13] In response to a request to publish the documents, the Secretary of State for Scotland told the UK House of Commons in 1989: "Data has been provided by the manufacturer of Nuvan 500 EC [dichlorvos] to the Veterinary Products Committee in support of the company's application for the product to be licensed for use as a medicine. Information in support of an application for a medicinal product licence is a matter of commercial confidentiality. The publication of such information and the tests on which it is based would be a matter for the company concerned."[14]

It is a similar story for all of the chemical case studies considered here—

governments are allowing private companies to hide behind commercial confidentiality clauses against the public interest.[15] Even work carried out by government agencies is either not published at all or is slipped out many years after the event.

Governments have sometimes given salmon farmers immunity from prosecution and virtual carte blanche to do as they please.[16] State-sponsored chemical pollution in Scottish salmon farming was such that the UK government ignored a 1994 recommendation by the Oslo and Paris Conventions for the Prevention of Marine Pollution (PARCOM) on best environmental practice for the reduction of inputs of potentially toxic chemicals from aquaculture use.[17] According to the Public Service Employees for Environmental Ethics, the British Columbia government has been in violation of its own Pesticide Control Act since 1995.[18] The Norwegian government breached its international obligations under the 1990 Hague Declaration when salmon farmers were allowed to increase their use of copper-based paints after the ban on TBT,[19] and the Chilean government allowed salmon farmers to use vast quantities of malachite green even after it was banned in 1995.[20]

Even the "polluter pays" principle has been turned on its head. In a recent case in Canada, the Sierra Legal Defence Fund revealed that salmon farming companies in British Columbia were reimbursed more than $1.7 million in fines. "Handing the fines back to the industry sends the wrong message," said Sierra Legal lawyer Tim Howard. "It encourages companies to knowingly violate their licences, and short-changes the taxpayer."[21]

These are not isolated cases, and they build up a picture of governments around the world bankrolling the expansion of sea cage salmon farming at the expense of both the environment and public health.

Chemical Case Studies

Chemicals used on salmon farms break down into four main groups:

- Antibiotics such as amoxicillin, oxytetracycline, oxolinic acid, sarafloxacin hydrochloride and sulphadiazine
- Artificial colourings such as canthaxanthin and astaxanthin
- Antiparasitics such as azamethiphos, cypermethrin, dichlorvos, emamectin benzoate, ivermectin and teflubenzuron
- Antifoulants such as TBT, copper and zinc-based paints

The use of antibiotics in salmon farming has been prevalent right from the beginning, and their use in aquaculture globally has grown to such an extent that resistance is now threatening human health as well as other marine species.[22] The Norwegian government's Directorate for Nature Management reported in 1999 that "during 1988–92, the mean annual consumption of antibacterial substances in Norwegian aquaculture was 29 tons, while traditional veterinary and human medicine used an average of ten to 25 tons a year in the same period. The aquaculture industry was thus the biggest contributor of antibacterial substances to the environment."[23] The directorate concluded that antibiotic use in salmon farming "may lead to an increasing number of resistant bacteria, and consequent treatment difficulties, in human medicine too."[24] Indeed, many of the chemicals used in salmon farming, such as oxytetracycline and amoxicillin, are also prescribed by doctors for flu and other infections.

Chile has taken over Norway's mantle as the world's number one antibiotic addict and is now using 75 times more antibiotics than Norway, which is taking its toll on both the environment and on the lucrative export market in Japan.[25] The abuse of antibiotics in aquaculture is a universal problem and not just confined to salmon farms—sea bass and trout farms, for example, are also showing similar signs of antibiotic resistance.[26]

Antibiotics and vaccines (which are a whole other story) are not considered in detail here, nor are the environmental and public health threats posed by PCBs, dioxins, dieldrin, toxaphene, chlordane, DDT and other organic contaminants that bioaccumulate "accidentally" via the fish feed. Instead, the chemicals considered are those used directly and deliberately on salmon farms. I am focusing on an artificial colouring, several antiparasitics, antifoulants and an antifungal.

- Canthaxanthin (E161g)
- Dichlorvos (Aquagard and Nuvan)
- Azamethiphos (Salmosan)
- Cypermethrin (Excis)
- Teflubenzuron (Calicide and Ektobann)
- Ivermectin (Ivomec)
- Emamectin benzoate (Slice)
- TBT (along with copper- and zinc-based paints)
- Malachite green

Artificial colourings and antiparasitic chemicals like emamectin benzoate, ivermectin and teflubenzuron are administered to farmed salmon via their feed. Antifoulants such as TBT or copper and zinc-based paints are coated onto salmon nets. For antiparasitic bath treatments like dichlorvos, azamethiphos and cypermethrin, farmed salmon are quite literally bathed in chemicals. The method of application is unsophisticated and in the bad old days involved leaking buckets and gloves with holes in them. A Scottish government report on fish farming explains: "Some chemicals, particularly antiparasitic and antifungal agents, are used as immersion treatments and the treatment solution is released to the water after use. Cage farmers add these chemicals directly to the cage, which is surrounded by a tarpaulin skirt or complete enclosure during treatment. Once a treatment is finished the skirt is removed and the chemicals are naturally flushed out . . . In most cases, these chemicals are used in such a way that they are directly flushed to the aquatic environment."[27]

Bath treatments are time-consuming (it might take as long as a week to treat an entire farm) and so toxic that they can cause stress and mortalities in the farmed salmon themselves[28]—not to mention their environmental and human health impacts. Consequently, in-feed treatments started to replace bath treatments in the late 1990s. Some farmers are even looking to inject their salmon with sea lice chemicals. A number of injectable anti-parasitics already exist for other species, and chemical companies are considering this form of delivery for salmon farming. Mass injections of farmed fish will cause some headaches. It is "impracticable to inject small fish," and even in large fish "there is a tendency for some of the injected material to leak back along the track of the needle." Moreover, "in the case of injection in any food animal, the implications for residues must be taken into account, especially at the site of injection."[29]

The following chemical case studies have some recurring themes including commercial confidentiality, political expediency, food safety and environment impact. The focus is predominantly on Scotland (and to a lesser extent Canada), largely because the information is more accessible. Due to language barriers, the world's number one and number two salmon farming nations, Chile and Norway, are pretty much out of bounds, although some information is slowly seeping out. Scotland is the third-largest salmon farming country in the world, producing 160,000 tonnes in 2003 (Norway and Chile produce about half a million tonnes each).[30] Scotland is also a microcosm of what is happening around the world. Some of the names of the chemicals may be different, but the issues are similar. The situation in the smaller salmon farming nations such as Australia, New Zealand, Ireland, the United States and Iceland may not be as bad as in Chile, Norway, Scotland, Canada and the Faroes (the big five)—but it's not necessarily much better either.

Before considering each chemical in turn, it is important to get a handle on the global consumption of chemicals by the salmon farming industry. Due to the reluctance of government agencies to divulge details of specific use of chemicals on salmon farms, it is difficult to give actual quantities used. Norway is the only salmon farming country that publishes annual figures.[31] The Scottish Environmental Protection Agency (SEPA) does "not hold comprehensive records of the total annual tonnages of chemicals used in the fish farming industry," and if records do exist, "commercial confidentiality" precludes their publication (Andy Rosie, SEPA, pers. comm.). Requests to government authorities in Australia, Chile, Canada, Ireland and New Zealand met with a similar response.

However, an extensive trawl of the literature gives a tantalizing and horrifying glimpse of the global chemical use in salmon farming.[32] An international survey published in 2000, for example, revealed that eleven compounds representing five pesticide types were being used on commercial salmon farms to kill sea lice: two organophosphates (dichlorvos and azamethiphos); three pyrethrin/pyrethroid compounds (pyrethrum, cypermethrin, deltamethrin); one oxidizing agent (hydrogen peroxide); three avermectins (ivermectin, emamectin benzoate and doramectin) and two benzoylphenyl ureas (teflubenzuron and diflubenzuron). The number of compounds available in any one country varied from nine (Norway) to six (Chile, United Kingdom) to four (Ireland, the Faroes, Canada) to two (United States). Dichlorvos, azamethiphos and cypermethrin were the most widely used compounds (5 countries) followed by hydrogen peroxide, ivermectin and emamectin benzoate (4 countries each), teflubenzuron (3 countries), diflubenzuron (2 countries), and deltamethrin, pyrethrum and doramectin (1 country each).[33] The situation in 2004 is probably very similar, although more countries have now licensed the use of both emamectin benzoate and teflubenzuron, and dichlorvos has been banned in several countries.

Calculating quantities of chemicals used is much more difficult, but there are some published surveys. Official government figures for annual chemical use by Scottish sea cage salmon farms were published in a private and confidential report in 1992: fourteen tonnes of formalin, 2,400 litres of vaccines, five tonnes of iodophors, 0.2 tonnes of furazolidone, five tonnes of ethoxyquin, 50 tonnes of dichlorvos, 0.3 tonnes of sulphadiazine and trimethroprim, two tonnes of chloramine T, ten tonnes of oxolinic acid, ten tonnes of oxytetracycline, 1.5 tonnes of malachite green, two tonnes of canthaxanthin, two tonnes of astaxanthin and 0.1g of methyltestosterone.[34]

The use of chemicals on salmon farms increased during the 1990s in line with increases in global salmon farming production. For example, in 1988 there was only one "medicine"—aqualinic powder—licensed by the UK's Veterinary Medicines Directorate for use on fish farms in the UK. By 2004 the UK government had issued over 30 licences including dichlorvos (1990), sulphatrim (1993), clamoxyl (1994), paramove (1995), azamethiphos (1996), sarafin (1997), cypermethrin (1999), teflubenzuron (1999), emamectin benzoate (2000) and pyceze (2003).[35] In Scotland, SEPA opened the floodgates to a wave of new chemicals in 1998 and by March 2004 it had issued over 1,000 licences to use the toxic treatments azamethiphos (282), cypermethrin (311), emamectin benzoate (211) and teflubenzuron (212).[36] A typical "discharge consent" issued by SEPA permits salmon farms to pump over 30 different chemical formulations including antiparasitics, antibiotics, antifoulants and disinfectants into the sea.

Canthaxanthin (E161g)—An artificial pink dye linked to eye defects

The most eye-catching chemicals used on salmon farms are the artificial pink dye canthaxanthin (E161g) and a related synthetic colouring, astaxanthin, which are added to the feed of farmed salmon for "colour finishing." Several major chemical companies produce canthaxanthin and astaxanthin, including agricultural behemoth Archer Daniels Midland and Swiss pharmaceutical firm Hoffmann-La Roche, which synthetically manufactured canthaxanthin from petrochemicals in its laboratories until 2002, when it sold its "vitamin and specialty chemicals division," Roche Vitamins, to Dutch chemical company DSM for $2.2 billion.[37]

Wild salmon get their pink and red colours naturally, mainly from eating krill, but salmon farmers can choose the colour of their farmed salmon with the help of the "SalmoFan," much as you pick the right shade of pink for your bathroom wall.[38] DSM, which sells canthaxanthin through its "nutritional products" line under the trade name Carophyll Red, now markets the "SalmoFan Lineal"—a colour-by-numbers slide ruler that illustrates every shade of pink from bubble-gum bright to dusty rose.[39]

Pink dyes can account for up to a third of all feed costs. For salmon farmers, though, it is a price worth paying. Strip farmed salmon of canthaxanthin and it is an unappetizing dirty grey in appearance. A consumer survey conducted by Hoffmann-La Roche found people put a premium on colour and are willing to pay much more for redder salmon. "Consumers perceive that redder salmon is equated to these characteristics: fresher, better flavour, higher quality and higher price . . . consumers felt that a salmon with a colour of

22–24 [on the "SalmoFan"] should be less expensive and a well coloured salmon, 33–34, would be the most expensive."[40]

Farmers using artificial colouring are deliberately confusing consumers, making money out of public ignorance, and endangering public health. Canthaxanthin has been liberally applied to farmed salmon since at least the 1980s despite fears that it can cause cancer. In 1990 the *Independent on Sunday* reported that levels of canthaxanthin in farmed salmon on sale in UK supermarkets exceeded safe levels,[41] while in 1992 the Scottish government admitted that "some concern has been expressed over the possible carcinogenicity of canthaxanthin, used to produce coloured flesh in farmed salmon."[42]

According to writer Linda Forristal: "There has been one reported death from aplastic anemia (failure of the bone marrow to manufacture red blood cells) attributed to the use of canthaxanthin as an oral tanning agent,"[43] and the scientific evidence of canthaxanthin's impact on the eye is extensive and dates back as early as 1987.[44] Hoffmann-La Roche knew even earlier than that. In 1986 the company wrote to a customer: "In investigations originally carried out in Canada, and more recently in Germany, a number of people who had been taking Canthaxanthin tanning preparations at high levels, for prolonged periods of time exhibited as a side effect a so far unexplained phenomenon which the authors describe as glistening, apparently crystalline deposits in the inner layer of the retina of the eye, in particular, around the macula. In some of the subjects investigated, sensitive ophthalmological tests revealed, slowing down of light-darkness adaptation of the eye, though the clinical significance of this remains to be fully determined. This functional disturbance is reported to regress on discontinuation of the Canthaxanthin tanning preparation. The deposits in the retina, however, have not been observed to regress, but remain in place without apparent impairment of vision, perception of color or field of vision."[45]

But while Hoffmann-La Roche considered it unsafe to take canthaxanthin via tanning pills, it still made millions selling canthaxanthin and astaxanthin to salmon farmers throughout the 1980s, 1990s and the current decade. Intense lobbying by Hoffmann-La Roche and the salmon farming industry has ensured that salmon farmers' "pink poison"[46] has been protected by the powers that be. In 1987 the UK government acted as a sponsor to Hoffmann-La Roche's scientific submission on canthaxanthin.[47] Scientific studies on health impacts of canthaxanthin, submitted by Hoffmann-La Roche to the World Health Organization and the European Commission, were classified as commercially confidential, and consumers were kept in the dark.[48]

In 1997 the European Commission's Scientific Committee on Food recognized the link between canthaxanthin intake and retinal problems but it was not until the EC's Scientific Committee on Animal Nutrition published a damning scientific opinion in April 2002 that the European Commission ordered Irish and Scottish salmon farmers to reduce levels of canthaxanthin by three to four times by December 2003.[49] The EC report set out all the scientific evidence linking canthaxanthin to retinal damage in the human eye and also revealed that since 1982 the levels of artificial colourings in the flesh of farmed salmon have more than trebled. Farmed salmon is getting pinker—chemical companies are getting richer.

The industry protested. "You have to consume enormous quantities of canthaxanthin before there is even the potential for damage to your eyesight," said Julie Edgar, the communications director at Scottish Quality Salmon (formerly the Scottish Salmon Growers Association).[50] She told the *Financial Times* that sales of Scottish farmed salmon were now threatened in France. "They judge our salmon a lot on its colour."[51]

Scottish and Irish salmon farmers did have the option to switch to astaxanthin, but the crux of the matter is cost: canthaxanthin is much cheaper than astaxanthin. Scottish salmon farmers estimate that the switch to astaxanthin "could increase the cost of finished feed from around £65 per tonne to

PHOTOGRAPHED BY MELISSA TATGE. REPRINTED BY PERMISSION OF ECOTRUST. WWW.ECOTRUST.ORG

Salmon farmers use the Salmofan™ to grade the degree of pink the flesh of their salmon attains through the use of canthaxanthin. Deeper shades of pink fetch higher prices in the marketplace.

The US Institute for Agriculture and Trade Policy's "Go Wild" campaign targets American consumers of farmed fish.

£80–£85 per tonne"—equivalent to £6 million per year.[52] Moreover, even the switch to astaxanthin is not without consumer concerns. When the EC's Scientific Committee on Animal Nutrition gave its opinion on canthaxanthin, it also recommended a food safety risk assessment of alternative artificial colourings, including astaxanthin.[53]

Of course, salmon farmers outside the European Union—in Norway, Chile, Canada, etc.—are exempt from the EC ruling and can continue to use canthaxanthin as long as they're not selling product in Europe. But if they do not see the sense of banning artificial colourings from their salmon altogether, they should at least come clean and inform consumers that salmon contain these dyes. The European Commission is now looking into labelling of both canthaxanthin and astaxanthin in farmed salmon (Beate Gminder, pers. comm.). The US already requires labelling, although supermarkets such as Safeway, Albertsons and Kroger have been accused of not following the law.[54]

Labelling would certainly have helped Linda Forristal. "It all started innocently with a delicious salmon dinner after a day's sightseeing in the Canadian Maritimes," she explains. "By midnight, I was itching inside and out, as if every blood vessel were dilated. Three days later, another salmon meal provoked an even worse reaction. That's when I discovered farmed salmon contain food dyes."[55]

Dichlorvos (Nuvan or Aquagard)—A carcinogenic organophosphate that is the skeleton in salmon farming's closet.

Dichlorvos (also known as DDVP) is a highly toxic organophosphate originally used by farmers and gardeners to control pests. Like many OP insecticides it disrupts the nervous and muscular system. Related to the military nerve gases developed in the Second World War, dichlorvos is one of the most toxic pesticides in the world. It was first registered in 1948 as an insecticide for use in agriculture and the home (trade names include Doom and Vapona). On salmon farms, the dichlorvos formulation is called either Nuvan or Aquagard (manufactured by Ciba Geigy—now known as Novartis). It was the first sea lice chemical to be used on salmon farms in the 1970s in Scotland and Norway and was also used extensively in Canada, Ireland, the Faroes and Chile. It is carcinogenic, mutagenic and hormone-disrupting—a so-called gender bender.

The National Poisons Unit describes dichlorvos as "toxic if swallowed," "very toxic by inhalation" and "toxic in contact with skin." It is treated as a "Red List" chemical under the European Commission's Dangerous Substances Directive [56] and is listed under the UK Poisons Regulations 1982, with harmful effects to humans from acute exposure and cumulative toxicity from repeated exposure to low doses.[57] The World Health Organization classifies it as "highly hazardous." Containers of dichlorvos must be labelled with the skull-and-crossbones warning sign and the following instructions for use: "Wear protective gloves, clean protective clothing, goggles, and a respirator of the organic-vapour type when handling this material. Avoid prolonged exposure to fumes. Wash hands and exposed skin after handling and before eating and bathe immediately after work. Keep the material out of the reach of children and well away from foodstuffs, animal feed and their

TOXIC

6

The makers of Nuvan, a product used to control sea lice, label their product with the skull-and-crossbone symbol for poison. Nuvan is 50% dichlorvos, a nerve toxin.

containers. Ensure that containers are tightly sealed, and stored and disposed of in such a way as to prevent accidental contact."[58]

Dr. Ted Needham (now boss of Heritage Salmon in Canada) first used Vapona fly-strips against sea lice in trials at Marine Harvest's Lochailort farm (then owned by Unilever, now by Nutreco) in Loch Ailort, Scotland. He chose dichlorvos because of its wide margin of error in killing sea lice and fish, and because the chemical was readily available for use in the agriculture sector. Needham first applied dichlorvos to salmon by dangling fly-strips in the water of the netpens.[59] (Nearly 25 years later, the same type of fly-strips Dr. Needham dangled in the water on a salmon farm were banned by the UK government after they were shown to cause cancer.[60])

Following these experiments in Loch Ailort, reports on slightly more professional experiments using buckets of dichlorvos were published by the Royal Society of Edinburgh in 1982.[61] Gordon Rae, a consultant for Scottish Quality Salmon up until 2003, when he died of cancer, described using dichlorvos in a *Fish Farmer* magazine article in 1979. Alongside is a photo of a fish farm worker pouring the carcinogenic chemicals into a sea pen with a plastic bucket. The worker has no gloves, mask or other safety equipment.[62]

Dichlorvos baths were a health hazard for the operator, the marine environment and the farmed salmon themselves. According to the Marine Conservation Society in 1988: "Not all farmers use tarpaulins or calculated quantities of Nuvan, resulting in considerable variability in the concentrations and over-use . . . One farm operator described the method of treatment to us as follows: 'Up to 30 cages can be treated together in an hour. No tarpaulin is used. The Nuvan is poured in at one end of the cage unit and is allowed to flow through with the tide, topping up the concentration at each cage. About 300–400 ml is used per cage. Treatment occurs four or five times per summer'. This would result in approximately 10.5 litres of Nuvan being used in one treatment of a group of 30 cages."[63] Dr. Ron Wootten of the Institute of Aquaculture at the University of Stirling told a House of Commons inquiry (1989–90) that dichlorvos was added to the water at slack tide, "which is frankly unacceptable as there is no containment and the stuff is just poured into the water."[64]

In 1989 the police, local schools and the Scottish Environmental Protection Agency were alerted after one company lost several litres of dichlorvos in Loch Sunart after they were blown into the water. An SOS went out for the missing bottles and a search party scoured over 90 kilometres of coastline. Dr. Paul Johnston, a toxins expert at London University and a

Greenpeace researcher, said: "[Nuvan] is in fact a nerve poison which could be deadly—especially if a child opened a bottle and took a deep sniff. It is environmental lunacy to keep this substance in a place where it can be blown into the water."[65] Following this embarrassing incident, Ciba Geigy, the chemical manufacturer of dichlorvos, was asked to provide better labelling—it was discovered that the Nuvan bottles' skull-and-crossbones warning labels washed off in salt water.[66]

The casual attitudes for using and storing dichlorvos are reflected in the quantities used. Although a Scottish government report detailing specific use of dichlorvos on each farm is still deemed private and confidential almost a decade after it was written,[67] it is possible to piece together some numbers from files obtained under the European Union's Freedom of Environmental Information regulations. Another private and confidential report from 1991 details how just two salmon farms operated by Marine Harvest had been using an average of nearly half a tonne (414 kg) of dichlorvos per year over the previous four years.[68] By 1989, Marine Harvest was using nearly three tonnes (2,920 kg) of Nuvan (50 percent dichlorvos) in Loch Ailort alone. A 1992 Scottish government report calculated dichlorvos use at a staggering 50 tonnes per year.[69]

Despite its widespread use across Scotland, dichlorvos was not granted a product licence by the UK government until June 1989. Even though the original risk assessments carried out in the 1970s remained private and confidential,[70] and, according to an article in *New Scientist*, Ciba Geigy admitted in 1988 that dichlorvos was not meant to be used on salmon farms,[71] the government caved in after intense lobbying by the industry, which basically said, "Let us use dichlorvos or we go under." Dr. Ron Wootten told the House of Commons inquiry into fish farming (1989–1990): "The crucial question as far as the industry is concerned is that without dichlorvos there would not be a salmon farming industry; there is absolutely no question of that."[72]

Ali Ross of the Whale and Dolphin Conservation Society argued against licensing at the inquiry. "You have to remember that it is now well over thirteen years since the chemical was first brought into use in the marine environment. It has been widely used and is now universally used by the industry and yet there are still substantial gaps in our knowledge on its environmental fate and potential impacts . . . We still have no published data from the Ciba-Geigy research that they submitted to the VPC [Veterinary Products Committee] and we have no published data from the DAFS research that is quoted in that paper; there has been no evidence that has been open to public scrutiny. As far as we are concerned, nothing has been presented yet that make us doubt the severe reservations we have about the continued and growing use of this chemical ... What causes great concern to us is that any judgment that is now made on the chemical is not just going to be a reflection of the scientific evidence that is presented supporting the case, but it is also going to be a reflection of the economic or political implications on the industry."

When the government issued the product licence, Ross's response was: "To give the impression that this stuff is environmentally friendly is immoral. If it wasn't for the fact that the substance was already being widely used by fish farmers, I am sure that it would never have been granted a licence. But as it is, environmental concerns have been quashed in favour of economic pressures."

Once it was legal, Scottish salmon farmers became so liberal in their application of dichlorvos that they used five times more than the combined consumption of the UK farming, pest control and household sectors. According to the Department of the Environment, in 1991 the annual usage of dichlorvos on Scottish salmon farms was ten to twenty tonnes compared to two to three tonnes for the entire terrestrial farming sector.[73] This was despite the fact that the UK government was committed to reducing inputs of dangerous chemicals, including dichlorvos, in the sea under international law. Dichlorvos was one of 30 "Red List" dangerous substances that the UK government had agreed to reduce by 50 percent by 1995.[74]

By the late 1980s sea lice had developed a resistance to dichlorvos.[75] This meant higher and higher doses were required, and it took days to administer a treatment. Mark Jones of the Fish Vet Group describes the cycle: "Nuvan, later known as Aquagard (dichlorvos) . . . only ever killed the larger, mobile stage pre-adult and adult lice. The safety margin for the fish was low, and with frequent repeated treatments required to remove successive waves of mobile stage lice, resistance developed quickly, necessitating increases in doses and/or exposure times in order to successfully remove the lice. This led to poor treatment success and many cases of repeated overexposure of fish leading to treatment kills. In addition, the concentrated product was dangerous for the handler, who risked long term health consequences in the event of repeated exposure."[76] In 1998, Marine Harvest Scotland admitted that a typical year class of farmed salmon could have as many as 22 treatments of dichlorvos.[77] The *Sunday Herald* reported in January 2002 that "according to some estimates, Scottish salmon farmers used a staggering 500 tonnes of dichlorvos through the 1970s, 1980s and 1990s."[78]

Dichlorvos was also used extensively across Norway. Trichlorfon (which breaks down into dichlorvos in seawater) was developed to kill sea lice in the mid-1970s before making way for the more potent Nuvan in the 1980s. Trichlorfon (manufactured by Bayer under the trade name Neguvon) was legally used in Norway until 1995, while dichlorvos was legal between 1987 and 1997.[79] Norwegian aquaculture used 3,488 kilograms of dichlorvos in 1989 and 3,588 kilograms in 1991, falling to 1,147 kilograms in 1994 and 161 kilograms in 1996.[80]

In the mid-1980s, Norwegian studies began to appear proving how toxic dichlorvos and trichlorfon were to shellfish such as lobsters, crabs, mussels, oysters and even other fish species such as herring.[81] The Canadian government also knew how dangerous dichlorvos was in the 1980s, yet it allowed salmon farmers in British Columbia and New Brunswick to use the substance throughout the 1980s and 1990s.[82] In 1988, a private and confidential Canadian report stated: "Unpublished work sponsored by Ciba-Geigy, the manufacturers of Nuvan, has shown larval lobsters to be lethally affected by dichlorvos at a concentration of 0.033 ppm after twenty hr."[83] Even farmed salmon can suffer OP poisoning from dichlorvos.[84]

The Canadian government discovered that the dichlorvos formulation Aquagard, which contains the solvent di-n-butyphthalate (DNP), was even more toxic to Atlantic salmon than dichlorvos alone. DNP is a really nasty piece of work and belongs to a class of compounds called phthalate acid esters (PAEs) that disrupt hormones and are on the priority list of most dangerous pollutants in Canada and the United States.[85] The industry calculated in 1993 that approximately eight tonnes of DNP had been released into the marine environment through aquaculture.[86]

Unpublished reports showing high levels of toxicity of dichlorvos date back over 40 years.[87] Information from the UN's Food and Agriculture Organization, for example, reveals that secret trials conducted since the 1960s detected dichlorvos residues in a wide range of foodstuffs including lettuce, rice, barley, wheat, cocoa beans, milk, cheese and meat.[88] Given the extensive use of dichlorvos in salmon farming, it is not at all surprising that Norwegian scientists detected dichlorvos in the flesh of farmed salmon in 1990.[89] Residues of dichlorvos were also found in farmed salmon on sale in supermarkets in the UK in the same year. Tests conducted by a Sunday newspaper made front-page headlines across Britain.[90]

Yet the Scottish, Irish, Norwegian and Canadian governments allowed salmon farmers to use dichlorvos throughout the 1980s and 1990s[91] until a scientific paper published in 1998 linked dichlorvos use on salmon farms to testicular cancer.[92] A team of scientists led by Professor Cecily Kelleher of National University of Ireland, Galway, found significant clusters of testicular cancer in salmon farming areas such as County Galway and Mayo. Researchers investigated the incidence of leukemia, lymphoma and testicular tumours in western Ireland between 1980 and 1990 and "found a significant increase in testicular tumours in agricultural workers other than farmers, albeit with very small numbers; this group comprised predominantly those engaged in fish farming."[93] Unfortunately, this Irish study is the only known study of its kind, and there are no plans to conduct further research (Prof. Cecily Kelleher, pers. comm.). Further studies are urgently required in other salmon farming countries where dichlorvos use has been widespread. The German Federal Environment Agency has listed trichlorfon (which breaks down into dichlorvos) as a potential endocrine disrupter that can cause mammary tumours and affect sperm and egg production, while a cluster of Down's syndrome children in Hungary was associated with consumption of fish excessively contaminated by trichlorfon.[94]

The whole sorry saga should have ended in 1999. Novartis claims it "voluntarily withdrew Aquagard from the market in November 1999 because it was superceded by Salmosan [active ingredient: azamethiphos], which is a clinically superior product."[95] However, the UK Department of Health did not finally ban dichlorvos until April 2002, after a damning report on its carcinogenicity in July 2001.[96] Yet extensive scientific evidence on the carcinogenicity, mutagenicity, and public health and environmental impact of dichlorvos had been available from the National Cancer Institute, World Health Organization, US National Toxicology Program and US Environmental Protection Agency since the early 1970s.[97]

Despite the UK ban, SEPA admitted in March 2004 that "around 20" Scottish salmon farmers were still permitted to use dichlorvos.[98] With staggering contempt for both the marine environment and public health, SEPA stated that "the process of going through registers to find all the Dichlorvos consents, so that they could be withdrawn, would deflect scarce manpower from the task of processing applications for the use of the current sea lice treatments."[99]

As early as 1988, Friends of the Earth Scotland warned of the cancer risk of dichlorvos use on salmon farms. FoE claimed a number of workers had been admitted to hospital with symptoms of Nuvan poisoning, such as severe nausea, headaches, dizziness and pupil dilation; some employees were being denied proper clothing, making the health risk even greater.[100] Reports from the National Poisons Unit show that between 1983 and 1990, 98 individual cases of poisoning involving dichlorvos were reported in the UK alone. It is not known how many of these cases relate to dichlorvos use on salmon farms, as information previously held by the UK government has been mysteriously "lost over time."[101] However, since salmon farmers used up to five times more dichlorvos than all other sectors, the industry is likely to account for a significant number of the 98 reported cases. The Health and Safety Executive's Pesticide Incident Appraisal Panel (1987–1993) did record a dichlorvos poisoning incident with the following sketchy details: "A 25 year old male worker became dizzy and nauseous whilst working in a hut used to prepare a sea lice treatment product Nuvan 50 EC."[102] Requests for further information have been refused by the UK government.

Occupational exposure is one thing, deliberate poisoning is another. In 1998 the Environmental Working Group published a report called "The English Patients," revealing how students at the University of Manchester had been treated with dichlorvos.[103] A California pesticide company hired a lab in England to conduct feeding trials using people to test the toxicity of dichlorvos. Hard-up students were actually paid to eat dichlorvos.[104]

Unsuspecting consumers of farmed salmon may have been paying for the same privilege for years. Norwegian, Scottish, Canadian, Chilean and Irish salmon farmers have all been unwitting guinea pigs in an authorized and highly irresponsible experiment lasting over three decades. These experiments could be coming back to haunt the salmon farming industry. According to the *Sunday Business Post,* a fish farm worker is suing a west of Ireland fish farm, claiming he developed testicular cancer while working with dichlorvos.[105]

In another legal action in Scotland, an ex-worker is claiming damages for health impacts he claims are related to an incident involving dichlorvos over a decade ago. On May 28, 1990, James Findlay says he was delousing fish, protected only by overalls and a small mask provided by his employer. Aquagard SLT, much more poisonous than dichlorvos alone, was added to a bucket of water and he was shown how to sluice it over the fish cages. As he did so, the bucket slipped in his hand and its contents went all over his head, face, shoulders and upper body. "I felt an immediate burning sensation and I wrenched the mask off, shoving my head forward to stop anything running into my mouth," recalled Findlay. "My eyes were burning, my shoulders seemed heavy. I was disorientated, I felt like I was floating. I knew I had to get ashore."[106] It took over three hours for Findlay to get to a hospital in Inverness, where he was given an antidote.

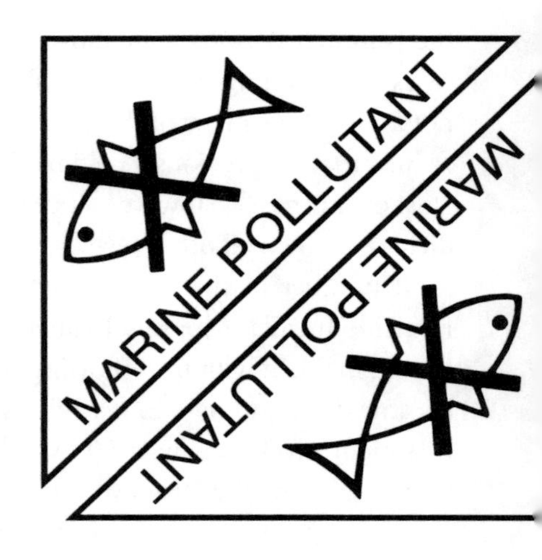

Azamethiphos is marked as a marine pollutant by its manufacturers. It replaced dichlorvos as a treatment for sea lice in the 1990s.

According to Findlay: "The medical prognosis is not exactly cheerful. My once very high IQ has been reduced by neuropsychologists to below average and I appear to have a form of autism when confronted with facts and figures and rapid analysis. I already have creeping paralysis and face total paralysis, further brain deterioration including dementia, numerous potential cancers, rapid ageing of the cells and, of course, early death. I live with depression, irritability, allergies, food intolerances, chronic pain and fatigue alongside tremors, panic attacks—you name it. Most painful of all, I have been told that I am sterile. I will never have the child or children I'd hoped for. It is just as well, because all the evidence suggests that if I did father a child the chances of genetic abnormalities would be incredibly high. Once I had a life of real promise. Is it any wonder that I expect someone, some company, some multinational to apologise and ultimately to pay for this?"

As Fidelma Cook of the *Mail on Sunday* concluded in her article on the case: "If successful in his claim, which could amount to tens of thousands of pounds, Mr. Findlay could open the floodgates to others who, despite worldwide studies backing their claims, have been denied any redress against the chemical industry, government and employees." Cook also pointed out that "farmers, Gulf War soldiers and even children who have been given head lice treatment have all been potential victims of the powerful pesticide which was first warned about back in the 1950s. Some opponents even suggest that the pesticide could have been responsible for BSE in cattle."[107]

Professor Malcolm Hooper, emeritus professor of medicinal chemistry at the University of Sunderland and chief scientific advisor to the Gulf Veterans Association told the *Mail on Sunday* that he had examined Findlay and was prepared to go to court on his behalf: "James is a victim of government compromise . . . It all comes down to commerce and cash and, as far as I'm concerned, the chemical industry is being protected by government." This skeleton in the closet looks set to rattle salmon farmers for years to come.

Azamethiphos (Salmosan)—An organophosphate nerve poison

When sea lice built up resistance to dichlorvos in the 1990s, farmers started flooding sea cages with a chemical ten times more toxic—azamethiphos, another organophosphate insecticide that affects the central nervous system.[108] Like dichlorvos, azamethiphos is manufactured by the Swiss-based chemical company Novartis (formerly Ciba Geigy) and sold for use on salmon as "Salmosan." It is authorized for use in Scotland, Norway, the Faroes, Canada and Chile.[109]

Also like dichlorvos, azamethiphos is poured directly into the marine environment, even though the manufacturer clearly warns against such use. For example, Novartis's safety data sheet for azamethiphos states: "Very toxic to fish." In capital letters it is marked as a "MARINE POLLUTANT." Under "Accidental Release Measures" the sheet warns: "Do not contaminate water courses or sewers."

There is nothing accidental about salmon farming's use of azamethiphos. The European Agency for the Evaluation of Medicinal Products stated in 1999: "The proposed use of azamethiphos in fish farming means that deliberate contamination of the environment will occur."[110] The Canadian government seemed to anticipate "accidental" releases of azamethiphos when it said: "The nature of the salmon aquaculture industry in southwest New Brunswick is such that many farms are in close proximity to each other and to areas of traditional lobster fisheries. Caution must be exercised to avoid an accidental release of significant quantities of these pesticides in sensitive areas."[111]

Azamethiphos was licensed for use on Norwegian salmon farms in 1994,[112] and official figures from Norway show that salmon farmers were using almost twice as much azamethiphos (738 kg) as dichlorvos (395 kg) by 1995.[113] The Norwegian Directorate for Nature Management reported in 1999 that "from 1993 to 1996 the use of the 'old' organophosphates dichlorvos and metriphonate (trichlorfon) plummeted—in excess of 90 percent reduction. In the same period, the consumption of azamethiphos doubled."[114]

Canadian salmon farmers were given "emergency authorization" to use azamethiphos in 1996, and it received a marketing authorization for use in Scotland in 1997.[115] SEPA started issuing licences for azamethiphos in 1998, despite admitting that "compared to dichlorvos, azamethiphos is more toxic to crustaceans" and that "there has been no short term study of the acute toxicity of azamethiphos to lobster larvae."[116] By March 2004 SEPA had issued 282 licences to use azamethiphos across Scotland.[117]

As in the case of dichlorvos, reports on the toxicity of azamethiphos were kept confidential by the chemical companies involved. For example, a 1992 report on the environmental assessment of azamethiphos is marked "Confidential to Ciba-Geigy Agriculture." It noted the chemical's toxicity to lobster larvae and also reported azamethiphos residues in the flesh of farmed salmon, but concluded bizarrely that: "Exposure of the general population to azamethiphos through treated fish should be negligible and should not constitute a health hazard. In spite of its toxicity, provided that the manufacturer's instructions are followed, azamethiphos should not constitute an undue hazard to those who are occupationally exposed."[118] Another report on "ecochemistry and ecotoxicity" of azamethiphos remains the private property of Ciba-Giegy Agriculture and has never been published.[119]

A risk assessment published in 1999 by the European Agency for the Evaluation of Medicinal Products indicated that "azamethiphos was mutagenic" but "overall, it was concluded that azamethiphos was not carcinogenic."[120] A slight improvement on the confirmed carcinogen dichlorvos perhaps, but not for the farmed salmon themselves, as the agency reported azamethiphos has a "very low therapeutic margin of safety in the target species, salmon." In other words, use too much of it and you end up killing your entire farmed stock. Mass mortalities from overdoses of both dichlorvos and azamethiphos have been reported across the industry—not surprising since farmed salmon are bathed in azamethiphos for up to an hour, and it takes up to a week to treat a whole farm. Nor is it a one-off hit. In 2000, Hydro Seafoods (now called Scottish Seafarms) "anticipated that 6–10 whole site treatments might be required per year."[121]

More recent studies have shown that even tiny concentrations of azamethiphos kill young lobsters.[122] The Department of Fisheries and Oceans in Canada reported in 2000 that "lobsters exposed to azamethiphos became agitated, often 'flopping' erratically around the exposure tank and became aggressive to other lobsters." Moreover, "they also seemed to lose control of their claws and eventually flipped onto their backs and died within hours." Signs of distress were recorded in adult lobsters at even ten percent of the recommended treatment concentration.[123] Since azamathiphos is a nerve poison, these findings are not altogether surprising.

The Scottish government has also admitted that azamethiphos (and cypermethrin) can cause shellfish poisoning and toxic algal blooms.[124] The link between chemicals such as azamethiphos and the stimulation of algal blooms was one of the driving forces behind an ongoing UK government study—"The Post-Authorisation Assessment of the Environmental Impact of Sea-Lice Treatments Used in Farmed Salmon."[125] Unfortunately this five-year study, which began in 1999, has still not been published.[126]

As well as the environmental impacts of azamethiphos, the direct human health effects have been investigated by the UK government. According to the Veterinary Products Committee in 1999: "Two areas of potential concern were noted. First, neurotoxicity studies did not meet modern standards. Secondly, no data on operator exposure were available." The most recent neurotoxicity study was carried out in 1991.[127] To plug the safety gaps, Ciba Geigy was asked to provide further information to the Advisory Committee on Pesticides in another clear-cut case of conducting risk assessments after the event. In 2001 the UK government recommended additional conditions on the use of azamethiphos.[128] By the time safety studies had been completed, sea lice were building up resistance to azamethiphos (and cross-resistance with dichlorvos), and though it was still considered a "very useful sealice treatment,"[129] it was already being phased out in favour of other chemical treatments such as cypermethrin, teflubenzuron and emamectin benzoate.

Cypermethrin (Excis)—A suspected "gender bender" and carcinogenic "neuro poison"

Cypermethrin (along with ivermectin) was the first in a new wave of chemicals replacing the organophosphates dichlorvos and azamethiphos.[130] It was developed by Shell (now American Cynamid) in the 1970s to control fleas, ticks, blowflies and lice on chickens, horses, cattle and sheep. Cypermethrin, a synthetic pyrethroid and neuropoison, has been used as a bath treatment on salmon farms since the mid-1990s. By 2002 it had been licensed for use in Norway, Ireland, Scotland, the Faroes, the United States and Chile under the trade name Excis (manufactured by Novartis). Two related compounds, deltamethrin (Alphamax) and high-cis cypermethrin (Betamax), are also available in Norway and the Faroes.[131]

There could be few worse replacements for the banned carcinogen dichlorvos. Synthetic pyrethroids are among the most potent pesticides and are hazardous to human health as well as to the environment.[132] According to the UK government, synthetic pyrethroids are "around 100 times more toxic to some elements of the aquatic environment" than organophosphates.[133] In a 1996 incident in Canada's Bay of Fundy, cypermethrin-contaminated effluent from a salmon farm was alleged to have caused the death of 44 tons of lobsters in a nearby lobster pond.[134] Pyrethroids have also been shown to be up to

1,000 times more toxic to fish than to mammals and birds.

Cypermethrin is a suspected hormone-disrupting compound—like dichlorvos and TBT—and can even affect reproduction in wild salmon.[135] Using it in sea cage salmon farms is difficult to defend on either environmental or public health grounds. Perhaps this explains why obtaining information is so difficult. As there was for dichlorvos, there has been a great deal of secrecy and complicity between government and chemical companies in securing the use of cypermethrin. Documents detailing its environmental impacts have either been kept "private and confidential" or were only published years after cypermethrin was first used.[136]

Cypermethrin, deltamethrin and permethrin were used extensively in "field trials" in Norway from as early as 1989 and used commercially across Norway from the mid-1990s.[137] According to the Norwegian Directorate for Nature Management, "cypermethrin was being increasingly used against salmon lice in 1996."[138] Norwegian salmon farmers used 215 kilograms of cypermethrin in 1997, falling to 69 kilograms in 2000 as sea lice became resistant.[139]

Field trials took place in Scotland as early as 1994 but were never published.[140] Trials also took place in Canada in 1996 but were not published until 2001.[141] Cypermethrin was not officially permitted in Scotland until 1999, yet in 1998 Wadbister Offshore Ltd. was fined £1,000 under the Control of Pollution Act when residues of cypermethrin were detected in mussels growing near Wadbister's salmon farms.[142] These incidents prompted the government to launch a program of screening for residues of cypermethrin in mussels and farmed salmon.

The illegal use of cypermethrin spread like a nasty rash across Scotland. "We were spraying this stuff all over the fish and inhaling it," whistle-blower Jackie Mackenzie told *The Observer* in 2000.[143] In the first signed testimony by a salmon farm worker alleging the illegal use of chemicals, Mackenzie described to me how Ardessie Salmon, then members of Scottish Quality Salmon, used Deosan Deosect (an illegal cypermethrin-based chemical): "The method used was to raise nets on the sea-cages to three metres and then surround pens with skirts and treat fish for one hour with a top-up after half an hour. When severe head shaking occurred in the fish pen, the treatment had to be aborted. We had a water pump affixed to a tank with sea-water and Deosan Deosect mix. It was then pumped into the cage via a sprinkler hose which dispersed the chemical. When treatment was finished we had to hand-ball the tarpaulin skirts back onto the boat-pontoon."[144]

Deosan Deosect is designed for use on chickens and horses. The manufacturer, Fort Dodge Animal Health (owned by American Home Products), classifies it as a marine pollutant. The label clearly warns: "Dangerous to fish and other aquatic life . . . Do not contaminate ponds, waterways or ditches with the product." A vet at Fort Dodge Animal Health told *The Observer*: "As far as marine life goes it is as toxic as you can get."[145] Subsequently a second whistle-blower stepped forward. "We used cypermethrin so many times I lost track," Jonathan Davis testified live on the BBC evening news.[146]

In 2001, empty cypermethrin containers washed up on a Shetland beach close to a salmon farm.[147] These were containers for Barricade, not Excis, the compound authorized for use by salmon farmers. A SEPA spokesperson told the *Shetland Times* that a fish farm could have saved thousands of pounds by using Barricade to treat sea lice instead of an approved product.[148] For salmon farmers wanting to cut costs, Barricade is readily available on the internet or via mail order.[149]

Writing in *The Scotsman,* BBC journalist Tom Morton, who lives in Shetland, pulled no punches: "Evidence emerged last week of callous, stupid malpractice within the industry when it comes to the use of illegal toxins. What happened was this: three empty tins of a substance called Barricade were dredged up from the seabed east of Shetland, amid a spread of salmon farms so intense you can barely see the water for cages. Barricade is used in the treatment of lice and fleas, on horses, and contains cypermethrin. It is an open secret that it offers fish farmers a cheap method of treating their stock—in a dangerous and wholly unscientific way—for sea lice. Saving thousands of pounds over approved methods. The Scottish Environmental Protection Agency was, understandably, annoyed. SEPA stipulates that cypermethrin is only used 'under the strictest conditions', including the full enclosure of the farm concerned and with monitoring to make sure the chemical disperses properly. But if you pick up some horse lice liquid, you can use as much as you like, where you like, whenever you like. And if it kills a few hundred lobsters and scallops, who cares? Well, I care. I'm sick of this arrogant assumption by salmon farmers that their right to profit comes before the environment I live in. They are poisoning our seas, and they are doing it with impunity. I did learn to like salmon. But, like most fisheries journalists I know, I won't be eating the farmed version again."[150]

On the other side of the Atlantic, salmon farmers in Maine were using legal and illegal versions of cypermethrin in 1997,[151] and SEPA reported in 1998 that "there is some illegal use of cypermethrin in Canada, often at night using high concentrations and no tarpaulin."[152]

Even when used legally, cypermethrin can cause fatal and sub-lethal impacts on a wide range of marine species including lobsters, crabs, mussels and salmon. Scientific research in Scotland, for example, has shown that high concentrations of cypermethrin cause valve closing in mussels[153]—not an ideal situation for a mussel farmer trying to earn a living next door to such a noxious neighbour. The same researchers also found impacts of cypermethrin on shore crabs.[154]

The environmental impacts of cypermethrin were recognized well before it was authorized for use on salmon farms.[155] Research carried out at the Canadian government biological station in St. Andrews, New Brunswick, during the 1970s, and later experiments on Prince Edward Island in the late 1980s, showed how lethal cypermethrin was to lobsters, shrimp and even salmon themselves.[156] Unfortunately, this research did not fully emerge until the late 1990s.[157] Confidential reports commissioned by Shell in the 1970s but never published also showed how toxic cypermethrin was to shrimp.[158]

In a 1996 review of both cypermethrin and azamethiphos, the Canadian Department of Fisheries and Oceans warned: "Unfortunately, these chemicals can be toxic to non-target marine organisms, including commercially valuable crustaceans such as lobster, and to other marine crustaceans that may be important to the coastal ecosystem." The DFO concluded that: "The distance travelled by a pesticide patch during the first two to four hours after release ranges from a few hundred to a few thousand metres, and hence may be carried through an adjacent fish farm . . . These findings have countered the often held belief by some government officials and industry that pesticides released into the marine environment of the Quoddy Region are instantaneously diluted. They have also reinforced the reality that many of the fish farms are sharing the water from adjacent farms on a regular basis and that knowledge of the circulation and dispersal patterns is valuable and necessary information."[159] A 1997 paper circulated internally by the Canadian government warned that cypermethrin was so toxic it would kill lobsters, crabs, prawns and other commercially important shellfish as well as sea lice.[160]

Canadian government researchers also showed that cypermethrin used on salmon farms in the Lower Bay of Fundy, New Brunswick, had "the potential to cause toxic effects over areas of hectares." Experiments at salmon farms in Deadmans Harbour, Letang Harbour and Black Bay in 1996 and 1997 showed "lethal effects" of cypermethrin at "extremely low concentrations"—some one to three orders of magnitude lower than the intended treatment concentrations. Nor was cypermethrin quickly diluted—"the plume retained its toxicity for substantial time periods after release." The researchers from the Department of Fisheries and Oceans and Environment Canada concluded that "since treatment of multiple cages is the operational norm, area-wide effects of cypermethrin on sensitive species cannot be discounted."[161]

Sadly, multiple treatments of cypermethrin do seem to be the operational norm, in Scotland at least. Hydro Seafoods (now called Scottish Seafarms), for example, states in a submission to SEPA that there is "a likely maximum of eight treatments per year" at one salmon farm in Scotland. And a private and confidential report reveals that a salmon farm in Loch Sunart used cypermethrin seven times in ten months.[162]

A Scottish scientific study, finally published in 2004 over a decade after the first trials took place, confirms the Canadian research. Copepods, organisms pivotal in virtually all pelagic food webs, died when exposed to cypermethrin at "considerably lower than the recommended sea lice treatment concentration." The researchers noted "animals showing increased activity in the form of erratic and frantic swimming, often swimming in small circles." Moreover, "sporadic twitching was also observed in animals prior to complete immobilization and has been reported previously in response to lobsters to cypermethrin exposure." The study concluded: "Consecutive treatments over several days at a salmon farm will introduce increasing levels of cypermethrin into the water column which may become entrained within a sea loch in localized currents. Thus, a cumulative impact of multiple treatments should not be discounted."[163] In other words, lethal plumes of cypermethrin are free to follow the currents, with the potential to kill sensitive species in their wake.

A SEPA survey published in February 2004 also found cypermethrin, teflubenzuron, ivermectin and emamectin benzoate in eleven percent of sediments tested under salmon cages. Cypermethrin was the biggest culprit, found in over a third of all positive samples. The highest concentration of cypermethrin was detected in Busta Voe in Shetland and was so high it exceeded SEPA's environmental standards. Other sites in Shetland at Collafirth, Bight of Cliffs and Ronas Voe also showed significant sediment contamination, with cypermethrin detected up to 100 metres away from the cages.[164]

Cypermethrin has also been shown to have significant effects on the reproduction and sense of smell of wild salmon. "The synthetic pyrethroid pesticide cypermethrin, a known contaminant of tributaries supporting spawning salmonid fish, had a significant sublethal impact upon the pheromonal mediated endocrine system in mature male Atlantic salmon," say researchers at the Centre for Environment, Fisheries and Aquaculture Science in England. "The results of the study suggest that low levels of cypermethrin in the aquatic environment may have a significant effect on Atlantic salmon populations through disruption of reproductive functions."[165]

Cypermethrin's capacity to affect the nervous system has been recorded in farmed salmon themselves. Grampian Pharmaceuticals (now owned by Novartis) warn in the safety data sheet for Excis (cypermethrin) that its use may cause "mild transient headshaking, flashing, increased jumping and uncoordinated swimming" in farmed salmon. Farm worker Jonathan Davis described a farmed salmon with "the shakes" at Ardessie Salmon: "We watched the swimming action of the fish and when we could see them starting to shake their heads, we stopped the treatment."[166]

Human health effects also warrant concern. Cypermethrin has long been considered a possible human carcinogen.[167] Studies on the carcinogenic and co-carcinogenic (tumour initiating and tumour promoting) properties of cypermethrin in mice revealed in 2002 that "cypermethrin possesses complete carcinogenic as well as tumour initiating and promoting potential."[168] The European Agency for the Evaluation of Medicinal Products warned in 1998 that "human occupational exposure to cypermethrin has been reported to cause transient paraesthesia on the face and other exposed areas of the body. It was considered that the paraesthesia was due to a spontaneous repetitive firing of the local sensory nerve endings, with thresholds temporarily lowered by the substances."[169] In other words, cypermethrin can cause numbness and a loss of feeling. Other reported symptoms of cypermethrin poisoning include abnormal facial sensations, dizziness, headache, nausea, itching, convulsions, burning and prolonged vomiting.[170]

That did not stop some Scottish salmon farmers from failing to provide workers with appropriate health and safety warnings. The safety instructions for the use of Deosan Deosect clearly state "Wear protective gloves, rubber boots and protective clothing," but Jonathan Davis, former farm worker, recalled that the gloves he and co-workers used "sometimes had holes in them as [the employer] didn't like to buy any new gloves. Even if the gloves were brand new and didn't have holes in, our hands got wet hand-balling the tarpaulin when using cypermethrin to treat the fish. The gloves always got soaking and at times my hands at the end of the working day looked like prunes . . . In terms of any health side-effects of using cypermethrin, there was a period of around two to three weeks when my nerve-endings in my fingers tingled when I had to reach for the rails around the cages. This happened in the period directly after treatment . . . On one occasion [the employer] opened a bottle of Deosect with his mouth. He used his teeth to pull the cap off. He obviously didn't value his own life let alone the health and safety of his employees."[171]

Davis's former co-worker Jackie Mackenzie is currently taking legal action over the use of cypermethrin, but even without pending legal action and studies confirming carcinogenicity in humans, cypermethrin's time may be up. Sea lice resistance to both cypermethrin and deltamethrin has already been reported in Norway and Scotland.[172] No worries for salmon farmers, though—they have plenty more toxic chemicals in the pipeline.

Teflubenzuron (Calicide)—A hazardous, wasteful and persistent marine pollutant

Teflubenzuron is a highly hazardous marine pollutant, lethal to shellfish in tiny doses, extremely persistent in the sediment under salmon cages and in the flesh of farmed salmon, and a suspected carcinogen. Hardly a suitable candidate for use on sea cage salmon farms, yet that is what salmon farmers reached for when they needed to replace dichlorvos and azamethiphos.

Teflubenzuron is a benzoylphenyl urea insecticide, initially introduced in 1984 to protect fruit, vegetables and cotton. By the 1990s, though, chemical resistance was already being reported in land-based pests,[173] so Nutreco (owners of Marine Harvest), in conjunction with the US chemical giant American Cynamid, developed teflubenzuron (trade name Calicide) for aquaculture. Sea cage fish farming is in danger of becoming a dumping ground for chemicals which are past their sell-by date on land.

Along with ivermectin and emamectin benzoate, teflubenzuron is administered to farmed salmon in their feed and not via a bath. It is now licensed for use in Norway, Scotland, Ireland and the Faroes and is available in both Canada and Chile on an emergency or trial basis.[174]

Norwegian salmon farmers started using both teflubenzuron (under the trade name Ektobann or Ectoban) and diflubenzuron (a related compound manufactured by Ewos [Mainstream] under the trade name Lepsidon) in 1996.[175] Official figures from the Norwegian Medical Depot and Norwegian School of Veterinary Science show that 610 kilograms of teflubenzuron were used in 1996, rising to 1,510 kilograms in 1997 and 1,334 kilograms in 1998.[176] Teflubenzuron is also listed by the Norwegian Directorate of Fisheries as equivalent to the chemical praziquantell,[177] which has been used in Norway since at least 1989. Norwegian salmon farms consumed over 1,500 kilograms throughout the 1990s.[178]

In 1998 Kurt Oddekalv of the Norwegian NGO Norges Miljøvernforbund claimed that teflubenzuron (and diflubenzuron) may cause cancer. The Norwegian media widely reported his fears that Norwegian farmed salmon contaminated with teflubenzuron were a health risk and unsafe to eat.[179] Following these revelations, the use of teflubenzuron in Norway plummeted five-fold in a single year to only 231 kilograms in 1999 and then to 28 kilograms in 2001.[180]

In Scotland, where teflubenzuron has been approved for use since 1999, fish may not be slaughtered for human consumption within seven days of treatment with the compound. In other countries the "withdrawal period"—the time it takes for drugs to flush out of the fish's system—is much longer.[181] Residues of diflubenzuron were found in Norwegian farmed salmon back in 1991, but it is not clear how widespread teflubenzuron contamination is.[182] There is not a shadow of doubt, however, over teflubenzuron contamination in sediments under salmon cages.[183]

The fundamental problem is that 90 to 95 percent of teflubenzuron ends up in the marine environment, mainly excreted through salmon feces.[184] And as the Scottish government's Marine Laboratory stated in 2001: "Once in the sediment, teflubenzuron could be available to the benthic community creating a possible passage into the food chain and the possibility of bio-accumulation."[185]

As well as being potentially carcinogenic to humans, teflubenzuron is highly hazardous to the marine environment, especially shellfish. The chemical works by inhibiting the formation of chitin, which is the predominant component of the exoskeleton of insects and crustacea. It is therefore highly toxic to species that undergo moulting at any stage in their life cycle, including lobsters, crabs and some zooplankton species. A safety data sheet from the manufacturers, Cynamid, warned in 1993 that teflubenzuron was "Dangerous for the environment," "Very toxic to aquatic organisms" and "May cause long term adverse effects in the aquatic environment." The Scottish Association for Marine Science warned back in 2001 of the lethal effects of teflubenzuron at tiny concentrations and over short periods: "The chemicals used to control sea lice are highly toxic to crustaceans, and are used by the salmon farming industry because of their efficacy at killing certain life stages of the parasitic copepods . . . Exposure of moulting stages to teflubenzuron and emamectin benzoate causes mortality and deformity at very low concentrations. Bioassay survivors displaying sublethal exposure effects do not generally recover indicating that sea lice chemicals may exert significant ecological effects at concentrations well below indicative LC50 values and after only brief exposures."[186] Diflubenzuron was also found to be toxic to crabs nearly a decade ago.[187]

Shellfish farmers have naturally objected to the use of teflubenzuron, arguing that a chemically dependent salmon farming industry is incompatible with shellfish farming.[188] Nor are we talking about small amounts of chemical. According to a 2003 application for a salmon farm in Loch Fyne, a "pessimistic" figure for the number of teflubenzuron treatments is three, with seven kilograms of Calicide per 100 tonnes of fish treated.[189] If 1,000 tonnes of farmed salmon were treated with Calicide three times a year, 210 kilograms of Calicide would be used, of which 90 percent (some 189 kg) would be discharged directly into the marine environment. Salmon farmers are not the only industry trying to earn a living in Loch Fyne—it is also home to the world renowned Loch Fyne Oysters.

Speaking to the *West Highland Free Press* in August 2000, scallop farmer David Oakes said: "Why was no work done on the effects of the chemical on scallops? The chemical affects the shells of the sea lice and it is likely it will affect shellfish as well, especially in the larval stages. There is evidence to show that the chemical is still to be found in the seabed six months after it was used."[190] Oakes gained this inside knowledge in 1999 when he was approached by a scientist working at a university in Scotland. The scientist had damning evidence concerning teflubenzuron, but wanted to remain anonymous. "Deep Trout" accused the Scottish government of failing to protect the marine environment and the shellfish farming industry.

Deep Trout's "Calicide Critique," widely circulated on the internet and submitted officially to the Scottish government and Nutreco, stated: "The prima facie evidence is that teflubenzuron will be highly toxic to shellfish. SEPA are therefore grossly ignorant of the range of species that will be directly affected by teflubenzuron. The lethal effects are: by prevention of growth in the Arthropoda; prevention of movement in the Annelida and death by starvation and internal damage in the Mollusca. There have been no studies of long term ecological effects of the use of teflubenzuron. They could be immense but have not been considered in its proposed use as Calicide."[191]

Deep Trout made a big splash in the Sunday papers. The *Sunday Herald* reported: "David Oakes, who requested the 'Deep Trout' report, has been denied access to the scientific evidence which led SEPA to accept that Calicide was safe. He was told that Trouw's (a subsidiary of Nutreco) commercial interests overrode the need for openness."[192] SEPA's pollution control specialist Andy Rosie told the *Sunday Herald* that "the claims made by Deep Trout are taken seriously." He admitted that SEPA's case was weakened by the fact that studies commissioned by Nutreco were still not available for peer-review. "At the very least the papers we have cited should be available. We will be negotiating with Nutreco to say that these papers should be in the public domain."

SEPA finally persuaded Nutreco to make some of the documents available for public inspection in February 2001, but they came in a straitjacket. In a letter to David Oakes, SEPA explained: "Nutreco retains copyright and intellectual property rights to most of these documents so SEPA is not able to provide you with copies. If you do wish to obtain a copy of any particular document, once you have viewed these, I would recommend you contact Nutreco directly."[193]

Subsequently, I visited SEPA offices to view a mountain of documents marked "Private and Confidential." I was given only a short time to peruse the material and was not allowed to make photocopies, but it soon became obvious why Nutreco might want to keep a lid firmly shut on these ecotoxicological studies. Different reports showed that teflubenzuron can persist in sediment for nearly two years at distances up to one kilometre away from the salmon cages; it can have "significant lethal effect on lobster juveniles fed salmon pellets containing as little as 0.5g feed additive per kg pellets"; "90 to 95 percent of the compound was excreted into the environment via feces"; "the highest concentrations were found 408 days after treatment"; teflubenzuron was still present 654 days after treatment; and it could bioaccumulate up through the food chain via filter-feeders such as scallops and mussels.[194]

The day after I viewed these documents, Nutreco's PR adviser Colin Ley rang me up. "How was your visit to SEPA yesterday?" he asked. "If there are any specific queries, please let us know—it would be much better to talk things through with us first before you make any public comments." Needless to say Nutreco read about the reports in the Sunday papers. "A controversial pesticide approved for use on 61 salmon farms in Scotland is classed as a highly toxic marine pollutant and can still be found in sediment on the sea bed nearly two years after use, according to documents revealed this week," reported the *Sunday Herald*.[195]

SEPA's role in authorizing teflubenzuron is typical of the manner in which governments in Norway, Canada, Chile and Ireland have allowed private profit to outweigh the public interest. SEPA's current policy on teflubenzuron was published in July 1999 and is based almost exclusively on Nutreco's unpublished private and confidential reports. The impact of teflubenzuron on crustacea such as lobsters is of primary concern, and SEPA's policy admits that teflubenzuron "is potentially highly toxic to any species which undergo moulting in their life cycle." As SEPA points out in the "environmental risk assessment": "This will therefore include some commercially important marine animals such as lobster, crab, shrimp and some zooplankton species."[196] In spite of this, SEPA began handing out licences to use teflubenzuron in 2000 and by March 2004 had issued 212.[197]

Meanwhile, the case against teflubenzuron is building all the time.[198] Little wonder then that salmon farmers want to bury the evidence. When a secret trial on its environmental impact was conducted in 1996 in the waters around the Isle of Skye, the first the locals knew about it was when their shellfish started dying. "We were unaware of the use of teflubenzuron until massive crab, prawn, squat lobster, and sea urchin deaths were observed in Lochbay," claims Aileen Robertson, who runs a diving centre in the area. "Scallop divers had to move to another sea loch, and the creel fisherman had to stop fishing. Even staff at the fish farm were alarmed to hear what was going on and gave us labels for the medicated food they had been given to use. We got the safety data, worked it out, and called the Scottish Environment Protection Agency. They had given consent for its sea trial with no public notification or advertisement. How do they get away with it?!"(Aileen Robertson, pers. comm.)

Ivermectin (Ivomec)—An illegally used neurotoxin and persistent marine pollutant

Like teflubenzuron, ivermectin is a persistent pollutant and in-feed treatment that seems to me wholly unsuitable for use on a sea cage salmon farm. Scientific studies show that it is acutely toxic to a range of marine life, it persists in the environment and it may potentially bioaccumulate in organisms and even humans. Ivermectin is a member of the avermectin group of chemicals, which are neurotoxins. It acts on the nervous system of invertebrate parasites, inhibiting nerve pulse transmission, resulting in their paralysis and death.

Manufactured by Merck Sharp Dohme as Ivomec, it has been used worldwide since the 1970s as a parasiticide for cattle, sheep, pigs, horses and dogs. Salmon farmers in Canada, Norway, Ireland and Scotland desperately (and illegally) resorted to ivermectin in the late 1980s and throughout the 1990s as sea lice became resistant to dichlorvos.[199] The Canadian government reported in 2000 that "the anti-parasitic agent ivermectin is prescribed by veterinarians and is used routinely in eastern Canada."[200]

However, ivermectin use produces severe side effects in animals. Scientific studies have shown toxicity in collie dogs and acute toxic syndrome in mammals, characterized by depression, mydriasis (excessive dilation of the eye's pupil), ataxia (loss of control of body movements), coma and death.[201] Nor is ivermectin a friend of the fish. Studies have shown that it can have direct toxic and pathological effects on farmed fish, causing uncoordinated swimming behaviour and respiratory problems.[202] As early as 1987, Irish and Canadian scientists showed that ivermectin caused increased mortality and listless behaviour in farmed salmon.[203] In 1993, researchers showed that Atlantic salmon treated with ivermectin became dark and lost their appetite, and their eyes rolled ventrally so lenses were no longer visible. Intestinal congestion was also recorded.[204]

That did not stop Canadians from giving thousands of farmed salmon a massive drug overdose. In January 2000 as many as 10,000 farmed salmon were killed at a farm in the Broughton Archipelego, BC, after an ivermectin treatment.[205] Dr. Joanne Constantine, fish health veterinarian with the BC Ministry of Agriculture, Fisheries and Food, concedes ivermectin has a "narrow margin of safety [for fish]" when it comes to calculating dosage.[206] Use too little and ivermectin does not do its job; use too much and it kills all your farmed stock.

Ivermectin also has fatal consequences for shellfish. Canadian government scientists concluded back in 1993 that ivermectin was lethal to shrimp at tiny concentrations.[207] Four nanograms of ivermectin per litre of water kills shrimp—that's 28 grams per 10,000 Olympic-sized swimming pools.[208] Such small doses can have an impact on consumers of contaminated farmed salmon as well. A study in 1995 by the Scottish government found that mussels bathed in a dilute solution of ivermectin accumulated the chemical to a concentration 750 times that in the water column. "In the worst scenarios, levels might be reached which affect the human embryo in the womb, the human baby through breast milk and the aged," said John Duffus of Heriot-Watt University in a report for the Association of Scottish Shellfish Growers. "I have considerable doubts as to the long-term safety of the use of ivermectin in the aquatic environment."[209]

The environmental impacts of ivermectin were known in the 1980s.[210] Like teflubenzuron, it is poorly absorbed by fish; a high percentage of the chemical is excreted via the feces and escapes to the marine environment. Residues of ivermectin in the flesh of farmed salmon and in the sediment under salmon cages are extremely persistent. The half-life is 90 to 240 days, which means it can take anywhere from three to eight months for half of the chemical in the sediment to decompose. Studies on ivermectin use on cattle have shown that it kills dung beetles in cow pats for up to four years after initial treatment. Professor Jean-Pierre Lumaret, from the University of Montpellier in France, found it produced "toxic and virtually indestructible" cow pats that are capable of killing up to 20,000 dung-eating insects a week.[211]

If ivermectin has such a toxic effect on dung beetles on land, it is no surprise to learn that it also kills sediment-dwelling organisms such as infaunal polychaetes living under salmon cages.[212] Research published in 1998 by Dr. Alistair Grant at the University of East Anglia showed ivermectin to be so toxic to marine life that he recommended a ban on its use on salmon farms. "It is clear that ivermectin is extremely toxic to some marine animals. In view of this, more data are urgently required regarding its toxicity and persistence in the field. It is difficult to justify its continued use until its environmental risks are understood more clearly."[213] Research by the Scottish government also found that ivermectin has a significant impact on lugworms—the marine equivalent of earthworms.[214] Dr. Ian Davies of the Scottish Office Marine Laboratory told *New Scientist* in 1998 that ivermectin use was so widespread in Scotland that "an area of between 10,000 and 20,000 square metres could be contaminated."[215] Dr. Davies later admitted in 2000, over a decade after the UK government knew it was being used, that "ivermectin can reach the marine environment via excretion from the bile, unabsorbed via the fish feces and by uneaten food pellets and has a strong affinity to lipid, soil and organic matter. Risk assessments have shown that ivermectin is likely to accumulate in the sediments and that the species therein would be more at risk than the species in the pelagic environment."[216]

Ivermectin has certainly left its mark on the Scottish seabed. In their survey of sediment under salmon cages published in February 2004, SEPA researchers found ivermectin residues across Scotland, even though farmers claim they stopped using ivermectin years ago.[217] Commenting on the findings, SEPA's lead aquaculture specialist said: "The study has detected low levels of unauthorized and unsuitable formulations, particularly ivermectin, and we are aware that an unscrupulous minority of companies have resorted to this in the past, mainly to save money."

Ivermectin contamination of sediment is one thing—the contamination of farmed salmon is another. Ivermectin lingers and leaves unappetizing residues in the flesh of farmed salmon.[218] So widespread was the illegal use of ivermectin that eleven percent of all Scottish farmed salmon tested in 1994 by the UK's Veterinary Medicines Directorate were contaminated with it.[219] No wonder supermarkets across Europe became wary of Scottish farmed salmon. In 1996 the German newspaper *Die Zeit* reported that Scottish farmed salmon was contaminated with ivermectin. The federal German Fish Research Agency "alleged that British salmon farms were getting around restrictions on marine use of the chemical by keeping small herds of sheep."[220] (As with cypermethrin, ivermectin formulations designed for use on terrestrial livestock could be bought cheaply over the counter in hardware stores.) Consumer pressure eventually led to some supermarket chains refusing to accept any salmon treated with ivermectin,[221] but as recently as 2001 ivermectin was routinely detected in samples of Scottish farmed salmon on sale in UK supermarkets.[222]

In spite of all the evidence about its toxicity and persistence, ivermectin has been used illegally to control sea lice on salmon farms across Scotland, Canada and Ireland since the 1980s.[223] In 1991 the manufacturer, Merck Sharp Dohme, wrote to Galway's University College Hospital: "This illegal use of ivermectin is neither encouraged nor condoned by the Company. MSD-AGVET will continue to oppose the use of ivermectin for treatment of louse infestations in salmon until all questions concerning safety and efficacy have been answered and the product is fully licensed."[224]

Also in 1991, a salmon farmer in Glencoe on the west coast of Scotland was caught using ivermectin by a filmmaker hiding in the bushes. According to the *Daily Telegraph*: "Fears that the substance is in use in parts of Britain emerged last week when a farm in Kinlochleven was raided by the Highland River Purification Board [later SEPA], after a tip-off by Friends of the Earth (Scotland). Acting on information provided by a farm employee that it was allegedly using ivermectin, Friends of the Earth filmed a fish worker not wearing goggles or protective clothing while feeding salmon. The worker later complained of difficulty with his vision. Users of the pesticide are warned to avoid contact with eyes."[225]

Friends of the Earth Scotland's Xanthe Jay said at the time of the incident: "It is difficult to imagine any other industry acting as irresponsibly as this and being so unregulated to get away with it. We accuse: the fish farmers who have been using ivermectin of being dangerously incompetent, the salmon industry who supposedly set quality standards of total negligence, and the statutory authorities charged with the control of the industry of providing nothing more than token safeguards. This shows the total inadequacy of the system of regulation of fish farming in Scotland. According to our calculations up to 6,000 tons of salmon contaminated with ivermectin could have reached the UK market."[226]

In 1993 another Scottish salmon farm company, Wester Ross Salmon, was fined for the unauthorized use of ivermectin in Loch Glencoul. Following raids by the enforcement agency, farmed salmon was found to be contaminated with ivermectin at a concentration twelve times above the detection limit. Wester Ross Salmon admitted feeding fish pellets containing ivermectin in a trial that ended in the death of thousands of fish. The company pleaded guilty to breach of the Control of Pollution Act and was fined a paltry £500.[227] Despite repeated government warnings, Scottish salmon farmers continued to treat the marine environment and public health with contempt. In 1995 two salmon farmers in Shetland were thrown out of the Shetland Seafood Quality Control scheme for using ivermectin illegally. Far from naming and shaming the companies concerned, however, "client confidentiality prevented the release of the salmon farms' identities."[228] According to the *Shetland Times*: "Sea-lice killer ivermectin is the skeleton in the salmon farming industry's cupboard. Everybody says everybody else is using it. Indeed, if you believe some people, all Shetland salmon farms use it."[229] Such widespread disdain for the law is by no means confined to the "unscrupulous minority." In 1996 the Scottish Salmon Growers Association (now Scottish Quality Salmon) issued guidelines on the use of ivermectin "premix for pigs" to treat farmed salmon.[230]

Canadian salmon farmers in British Columbia were also caught out using ivermectin illegally. Freedom of information requests by the Sierra Legal Defence Fund revealed that in 1997 alone, 107 kilograms of ivermectin were dumped into the ocean at salmon farms on the west coast of Vancouver Island.[231]

Scottish salmon farmers' thirst for this illicit "jungle juice," as it is known in the trade, led to cattle farmers in Shetland complaining that stocks of ivermectin were running dry (Anon, pers. comm.). For a brief time between 1996 and 1999, SEPA allowed a small number of Scottish salmon farmers (about 30) to use ivermectin legally.[232] But in February 1999 the Secretary of State for Scotland effectively shut the door on the legal use of ivermectin by ordering a public inquiry.[233] Salmon farmers subsequently pulled the plug on the chemical when they realized such a toxic product would not stand close scrutiny.[234]

Some Scottish salmon farmers, however, did not let the law or public health concerns get in their way. One such operator was Ardessie Salmon, which came under the spotlight in a front-page exposé published by *The Observer* newspaper in April 2000.[235] Jackie Mackenzie told me in a signed testimony: "I used ivermectin on smolts on numerous occasions. We added ivermectin to smolt food by adding a given amount to water and then, using a knapsack sprayer, we coated the food in a concrete mixer. We then transferred the treated food back into bags and then hand fed over the weekend so the Scottish Environment Protection Agency could not do an unsuspected inspection."[236] Following the revelations of Mackenzie and another colleague, Ardessie Salmon was thrown out of Scottish Quality Salmon, which also withdrew its Tartan Quality Mark from the farm's tainted products.[237]

As recently as 2002, yet another Scottish salmon farmer was caught using ivermectin, fined a record £6,000 and forced to withdraw from the Shetland Seafood Quality Control scheme.[238] Speaking in court afterwards, SEPA's water pollution officer in Shetland, Dave Okill, lamented: "I feel disappointed that this part of the industry has seen fit to use an unlicensed and uncontrolled chemical in this way. I am also disappointed that the Shetland Salmon Farmers Association has failed to convince this section of the industry that they have environmental responsibilities, and that they should recognize those responsibilities. This firm has obviously not just broken the law, it also has broken the Code of Best Practice as issued by the Shetland Salmon Farmers' Association, and agreed to by its members. We have in the last five years taken two reports to the Procurator Fiscal about the illegal use of chemicals in the industry. I think I would be naïve to believe that those two reports related solely to the only two times that illegal chemicals had been used."[239]

Emamectin benzoate (Slice)—A marine pollutant toxic to fish, birds, mammals and aquatic invertebrates

Salmon farmers have merely moved on from one toxic chemical—ivermectin—to another—emamectin. The use of emamectin benzoate is legal, but whether or not it is any better is debatable. Emamectin benzoate is a semi-synthetic avermectin insecticide, closely related to ivermectin. It has been sold for agricultural pest control in edible plant crops since the 1970s, but was only added to salmon farmers' chemical arsenal in the 1990s. The safety data sheet for Proclaim (active ingredient: emamectin benzoate) warns: "The pesticide is toxic to fish, birds, mammals and aquatic invertebrates. Do not apply directly to water or to areas where surface water is present, or to intertidal areas below the high water mark."[240]

Against this safety advice, emamectin benzoate is now widely used on sea cage salmon farms under the trade name Slice, manufactured by the US pharmaceutical giant Schering Plough. Like teflubenzuron, it is administered to farmed salmon as a premix coated on fish feed. Slice is licensed for use or is being used on a trial basis on salmon farms all over the world including Scotland, Chile, Norway, Ireland, Iceland, the Faroes and Canada.[241] Secret field trials took place in Scotland as early as 1994 under the code name SCH5844. Further trials in Scotland took place in 1997 but were not published until 2000. Field trials also took place in Canada in 1998.[242]

With sea lice resistant to dichlorvos, azamethiphos and cypermethrin, emamectin benzoate is the current chemical weapon of choice in the war on these parasites. However, like ivermectin, emamectin benzoate is no "magic bullet."[243] And like dichlorvos, azamethiphos or cypermethrin, it can produce after-effects. Schering Plough's catchy sales slogan is "Slice Kills Lice." Sadly, Slice may also kill other marine life. Unsurprisingly for a toxic chemical labelled by its manufacturer as a marine pollutant, emamectin benzoate pollutes the marine environment. However, many salmon farmers seem to have been more concerned about its efficacy than its ecotoxicology.[244]

Field trials on farmed salmon showed "signs that were compatible with toxicosis." One fish was so drunk on emamectin benzoate that it "swam with its snout out of the water and appeared to be losing equilibrium." Other conditions reported in the study included scale loss, peritoneal adhesions, visceral melaniation, multi-focal gill lamellar fusion.[245] Another study reported lethargy, dark coloration and loss of appetite in both salmon and trout treated with emamactin benzoate. One farmed salmon was so anorexic it died from "focal necrosis, ceroid accumulation in the spleen and melanin accumulation in the kidney." Other side effects included skin lesions and erosion of the pectoral fins, nose and mandibles. Corneal edema and cataracts were recorded in almost half the salmon treated with medium doses of emamectin benzoate. Another blind-drunk salmon on a high dose of emamectin benzoate "was observed to roll onto its side at intervals, apparently unable to maintain an upright position."[246]

Emamectin benzoate is lethal to a wide range of crustacea (not just sea lice) at very low levels. Studies by SEPA show that the small mysid shrimp is poisoned by emamectin benzoate at concentrations equivalent to only half a drop in an Olympic-sized swimming pool.[247] Research by the Scottish Association for Marine Science also concluded that both emamectin benzoate and teflubenzuron cause "mortality and deformities at very low concentrations" in non-target planktonic copepods.[248] Another Scottish study reported in 2003 that experiments showed emamectin benzoate significantly reduced moulting success, reduced fecundity and caused deformities in copepods.[249] Copepods—the microscopic aquatic equivalent of ladybirds and beetles—are vital to the health of the ecosystem and are an important food supply for wild salmon and other fish species.

Species much larger than copepods are affected by emamectin benzoate as well. Canadian government researchers stated in a paper published in 2002 that emamectin benzoate can cause premature moulting, failure to reproduce and death in lobsters on the east coast of Canada. The studies "confirm the molt-producing effect of emamectin benzoate on female American lobster." Furthermore, the "results provide conclusive proof that emamectin benzoate is disrupting the endocrine system that controls molting in the American lobster. The results are the first example of a crustacean molting prematurely in response to chemical exposure, the first example of an arthropod molting in response to an avermectin, and the first report that GABAergic pesticides can induce proecysis in crustaceans."[250] Translated into English—those lobsters won't be raising a family in a hurry. Similar impacts are predicted for species closely related to lobsters including prawns, crabs, and shrimp. Emamectin benzoate is also highly toxic to the northern bobwhite quail and the mallard duck.[251]

Emamectin benzoate, like ivermectin, is hard to shake off and sticks like glue to sediments. Residues have been found in soil, water and crops growing in contaminated soil,[252] and it leaches into the marine environment months after a salmon farm treatment.[253] Unsurprisingly, it is appearing in the seabed under salmon cages. In its survey of sediments sampled under salmon cages, mentioned earlier, SEPA detected residues of emamectin benzoate at three times more than the "monitoring trigger value within 25 m of the cage edges," and also detected it in the flesh of farmed salmon, but "well below the maximum safe limit for human consumption."[254]

It is so persistent that in Norway and the Faroes the withdrawal period for farmed salmon treated with Slice is 120 days before harvesting. This does little to prevent contamination of other foodstuffs such as wild shellfish. For example, emamectin benzoate has recently been found in wild scallops near a salmon farm in Maine, USA. Tests carried out in Cobscook Bay found emamectin benzoate contamination at three times the food safety limit set by the US Environmental Protection Agency. The National Environmental Law Centre (NELC) says the discovery warrants a warning to harvesters and seafood consumers—the area is a commercial fishery for scallops, lobster and pollock.[255]

The NELC has also called on the Maine Department of Environmental Protection to conduct rigorous, comprehensive monitoring of non-target marine organisms for residues of drugs and other chemicals used at salmon farms (Josh Kratka, pers. comm.). Sediment scavengers such as prawns, shrimps and crabs are at risk and may be contaminated with emamectin benzoate.

The chemical manufacturer, Schering Plough, has presumably known about the contamination of shellfish for over five years. Confidential research dating back to 1999 but never officially published reveals that emamectin benzoate was detected in mussels up to 100 metres from salmon cages one week after treatment.[256] As far as the manufacturers are concerned, Slice is still safe, but Schering Plough has apparently not been anxious to send papers on environmental impact to journals for peer review (John McHenery, pers. comm.), and the vast majority remain private and confidential.[257] Some documents on the environmental impacts of Slice are available on the internet, far from any peer-reviewed scientific journals.[258]

Schering Plough's record in other areas does little to inspire consumer confidence. In 2002 the company was given a whopping US$500 million fine—the largest ever—by the US Food and Drug Administration for significant violations with respect to their facilities, manufacturing, quality assurance, equipment, laboratories, and packaging and labelling.[259] In October 2003 Schering Plough was fined again—this time a US$1 million civil fine to settle federal regulators' allegations that it illegally revealed financial information. With plummeting sales, Schering Plough is described as "one of the most beleaguered companies in the US pharmaceutical industry."[260]

Faced with a mountain of scientific evidence on the dangers of emamectin benzoate, Scottish and Canadian governments worked to approve Slice's use on salmon farms—yet another example of government putting salmon farmers and chemical companies first and shellfish and food safety second.

According to a report on the CTV television network in May 2002, Health Canada had "badly abused" the emergency drug-release program to allow salmon farmers across Canada to use Slice. The chemical had been used "more than 770 times in the past year alone." The *Globe and Mail* reported that, according to documents obtained by CTV, "several toxicology studies of Slice (emamectin benzoate) done on rats, dogs and rabbits show that when ingested at high doses the drug can cause side effects such as tremors, spine and brain degeneration and muscle atrophy."[261] Earlier, in October 2000, the *Ottawa Citizen* reported that the Canadian government was allowing salmon farmers to "side-step drug ban."[262] In an assessment report in June 2001 the government's own agency, Health Canada, criticized the Canadian Food Inspection Agency for failing to test for residues of emamectin benzoate in farmed salmon even though it knew full well that the chemical represented over one in three (38 percent) drug prescriptions in the Atlantic area.[263]

The organization Public Service Employees for Environmental Ethics (PSE) believes the British Columbia government is guilty of the same chemical corruption with its "aggressive treatment" strategy to kill sea lice. In March 2003 the PSE accused the BC government of being in violation of the provincial Pesticide Control Act since 1995. "Why is the government ignoring their own law?" asked Mike Romaine, executive director of the PSE. "Laws such as the Pesticide Control Act are in place to protect the environment and allow for proper public review. The members of the PSE demand that the provincial government stop allowing the use of these dangerous chemicals in the marine environment without a permit."[264]

In a background briefing on emamectin benzoate the PSE stated that BC has "turned a blind eye to emerging concerns" and has been "sympathetic to industry with respect to the enforcement of regulations, monitoring and management . . . Government is now looking to fast-track the use of pesticides or drugs that are known to have environmental concerns and to ignore their own laws with respect to enabling concerns to be aired through a review process. Documents show that when sea lice became a problem in 1995, the industry attempted to use unregistered pesticides to bypass the BC Pesticide Control Act."[265] Speaking on CBC-TV, Romaine accused the Canadian government of ignoring the dangers in its zeal to keep the aquaculture industry afloat. "They've got a bad situation and want to find a silver bullet," he said. "They're rushing to do everything at the expense of regulation, sound science and democratic decision-making."[266]

Canadian salmon farmers have some way to go before they reach the murky depths plumbed in Scotland (illegal chemicals have been used there since 1976), but they are learning fast. Not to be outdone, the Scottish government also allowed salmon farmers to use emamectin benzoate before it had properly dealt with the applications. In February 2002 SEPA fast-tracked a stream of applications to use emamectin benzoate without carrying out appropriate risk assessments.[267] Its decision to short-circuit the regulatory process is yet another example of the government bending over backward for the salmon farming industry. The timing could not have been worse for shellfish farmers—it coincided exactly with the time the free-swimming larval stages of crustacea and other marine life were spawning. SEPA's policy on emamectin benzoate (dated 1999) is now five years out of date and fails to take in much new scientific information.[268]

Scottish salmon farmers have also been caught using emamectin benzoate contrary to the manufacturer's instructions. The UK's Veterinary Medicines Directorate "is concerned that, if the product becomes less effective, this could in turn lead to more frequent treatments needing to be used, which will increase the overall burden on the environment."[269] Misuse may lead to sea lice developing resistance more quickly, and Schering Plough has every reason to worry about this. Resistance to emamectin benzoate in land-based agriculture occurred a decade ago,[270] and the company clearly wants to protect its $10 million investment. Canadian government scientists are already engaged in a study on sea lice resistance to emamectin benzoate, with results to be published in the summer of 2004 in *Pest Management Science* (John Burka, pers. comm.). It is only a matter of time before salmon farmers encounter the same sea lice resistance to emamectin benzoate that they have experienced with dichlorvos, azamethiphos and cypermethrin. Slice may soon be more of a blank than a magic bullet for sea lice.[271]

TBT—A highly toxic antifoulant paint used throughout the 1980s

Tributyltin (TBT), according to Professor Edward Goldberg of the Scripps Institute of Oceanography in the United States, is "one of the most toxic substances deliberately introduced into our natural waters."[272] It was first used by ships and yachts as an antifoulant paint in the mid-1960s and was quickly adopted by the salmon farming industry in the 1970s. Despite increasing evidence that TBT caused reproductive failure and growth abnormalities in shellfish, it was used to coat salmon farm nets all over the world. Scottish shellfish farmers complained throughout the 1970s and 1980s that TBT contamination was forcing them out of business. TBT was called "a crime against nature and the shellfish farming industry."[273]

Scientific evidence showing the damaging effects of TBT on oysters was published by the Fisheries Research Board of Canada as far back as 1967.[274] Further research showed TBT was lethal to mussels, oysters and scallops and was so toxic it could change the sex of dogwhelks and snails.[275]

The French government banned TBT in 1982 to save its oyster industry, but the Canadian, Norwegian and Scottish governments were prepared to sacrifice the shellfish industry for the sake of salmon farming. The Scottish government continued to sanction the use of TBT on salmon farms even though it knew scallops and oysters on sale to the general public were contaminated with the chemical.[276]

In Scotland, shellfish farmer Allan Berry launched the "Ban TBT Now" campaign in 1984 after noticing his oyster shells were being affected (there are echoes here of Rachel Carson's revelations regarding DDT and its impact on birds' eggshells). "TBT was the most potent molluscicide known and it was wrecking our oyster production," says Berry. "Instead of being long and thin they had small, very thick shells—a bit like walnuts—and the meat was not developing inside them. Salmon farming poisoned us out of business—not just me, a whole ruddy industry." But the political might of the Scottish salmon farming industry withheld a ban until 1987. "By this point all the west coast shellfish growers were being affected more or less, but they were small operators, crofters and part-timers who didn't know how to argue with government. The salmon farmers had suits, degrees, huge Range Rovers and quoted huge sums of earnings," says Berry.[277]

It was food safety issues in the economically more important salmon farming sector that finally forced governments to act. Researchers from the US National Marine Fisheries Service in Alaska found that farmed salmon could absorb TBT from the antifoulant paint on the cages.[278] Their study showed TBT contamination in eleven out of fifteen samples of farmed Pacific chinook salmon bought from markets in Seattle and Portland—55 to 76 percent of the TBT was still present after cooking.[279] Another American study showed that human red blood cells were extremely sensitive to TBT, with tiny concentrations inducing membrane breakdown.[280]

In 1986 the US Food and Drug Administration declared farmed salmon containing TBT to be unfit for human consumption, but other countries were slower to act. The UK government confirmed in 1986 that farmed Atlantic salmon on sale for human consumption was contaminated with TBT, but did not publish the information until 1987.[281]

Andrew Lees of Friends of the Earth urged the UK government to "ban TBT immediately. The Environment Protection Agency found serious defects in safety tests used to obtain approvals in the United States for products containing TBT. There is mounting evidence of hazards to the environment and potential threats to human health. TBT is on trial and our verdict is guilty. The Government should arrange for the testing of all seafood products, including oysters and salmon, derived from waters likely to be polluted with TBT. Contaminated products should not be marketed and any products in the supply chain should be withdrawn from sale."[282]

In the wake of the bad publicity surrounding TBT contamination of farmed salmon, major supermarkets issued an ultimatum to salmon farmers—stop using TBT or we stop selling farmed salmon.[283] Less than a week later, on February 25, 1987, the UK government finally banned TBT use on salmon farms.[284] It took another year for the Scottish government to publish studies on the effects of TBT on oysters.[285] Norway was even slower to act and banned TBT in 1990. In 1992 the Scottish government also conceded that TBT was "implicated as growth stimulants for toxic 'red tide' producing dinoflagellates in studies elsewhere in Europe,"[286] joining dichlorvos, azamethiphos and cypermethrin as implicated in the stimulation of toxic algal blooms.[287]

Some Scottish salmon farmers simply stockpiled TBT before the ban came into force. The *West Highland Free Press* reported in April 1987 that "one large-scale salmon farmer in the north-west mainland is at present stockpiling supplies of anti-fouling substances which contain TBT, possibly with a view to offering them for future use."[288] In May the *Shetland Times* also reported "quite a rush" on TBT before it became illegal. Commenting on a sign in a local shop urging salmon farmers to stock up before the ban, Douglas Smith, director of environmental health at Shetland Isles Council, said: "Considering what's happened in the recent past and considering that the ban takes effect very shortly this does seem slightly less than responsible."[289]

The ban on TBT left other salmon farmers with a toxic waste disposal problem. One of them, Laurence Anderson of Sunnyside in the Shetland Islands, "dumped three barrels of toxic anti-fouling compound over a cliff and ordered three more barrels to be dumped in the sea." After Anderson was fined £600, defence agent Steve Leeman said that his client had wanted to dispose of TBT "because it was about to be banned but he had not realised how dangerous it was."[290] A subsequent scientific assessment by the Scottish government reported on this illegal dumping's impact on dogwhelks and limpets.[291]

A decade after it was banned in Scotland, the government eventually "published" a report on the effects of TBT, but it is marked "Private and Confidential."[292] Nearly twenty years after TBT was found in farmed Scottish salmon, the European Commission's Scientific Committee on Food is still preparing a scientific opinion on the risk assessment of TBT in fishery products. The TBT issue also raises questions about all the other chemicals licensed for use on salmon farms. Just after the TBT ban on Scottish salmon farms in 1987, an editorial in the *Glasgow Herald* asked: "The remaining worry about TBT is this: if it was examined, as it presumably was, and passed fit for introduction to the environment by the appropriate Government agency, how many other substances of a similarly powerful toxicity have received the same approval?"[293]

For a start you can add copper- and zinc-based paints—the less-effective biocides that replaced TBT—to the long list that already includes canthaxanthin, dichlorvos, azamethiphos, cypermethrin, teflubenzuron, ivermectin and emamectin benzoate. The salmon farming industry saw copper as "safer than tin," but safety is all relative.[294] Research by the Norwegian government as far back as 1985 (even before the ban on TBT) shows that copper is acutely poisonous to many marine organisms and that bioaccumulation occurs in algae, oysters, mussels and crabs.[295]

The safety data sheet for Aqua Net, one of the copper-based paints used by salmon farms, states: "Do not empty in sewers or other water drains" as it contains a substance classified as "toxic for water-living organisms" and "may cause unwanted long-term effects in the water environment." Canadian marine biologist Alexandra Morton reports that many of the netpens in the Broughton Archipelago of British Columbia are daubed red with copper. They have been painted with Flexgard XI—active ingredient: 26.5 percent cuprous oxide. The label for this paint sports a skull and crossbones and a "notice to user" that says: "Product to be used only in accordance with the directions. Toxic to aquatic organisms. Do not contaminate water. Do not allow chips or dust generated during paint removal to enter water."[296]

According to the Norwegian Directorate for Nature Management, most of the copper emissions from salmon farming (80 to 90 percent) take place as diffuse releases from the actual nets as the copper dissolves in the water.[297] The way salmon farmers clean their cages accounts for the rest. Nets are often washed on beaches. Washing not only removes fouling organisms such as barnacles, mussels and seaweed, but also copper and zinc (and TBT when it was used), which are then flushed into the sea. Fisheries Information Service (FIS) has reported that "local copper pollution has occurred near fish cage maintenance facilities in Norway. Discoloration of beaches has been noted and increased copper concentrations in sediments have been found several places."[298]

The quantities of antifoulant used on salmon farms are alarming. For example, an Irish salmon farm predicts that it will use "7,364 litres of copper-based antifoulant per year" for 1,500 tonnes of farmed salmon.[299] In Norway, 529 tonnes of copper were released to water by the boating, mining, industrial and salmon farming sectors in 1985, rising to 647 tonnes in 1996. But while the industrial sector's copper use fell six-fold during that time, the salmon farming industry's use increased four-fold. Such increases were in contravention of the 1990 Hague Declaration and the 1995 Esbjerg Declaration, which committed the Norwegian government to reduce and then cease emissions of copper by 2020. In 1997 the Norwegian Parliament ordered that "emissions should be substantially reduced by 2010,"[300] and in 2001 the Norwegian Pollution Control Authority finally put forward a proposal to ban emissions of copper-based paints from salmon cages,[301] though the ban may come too late to save some beaches.

Some of the copper released by salmon farms is bound in organic or inorganic compounds that gradually sink to the sea floor. Over time this leads to a rise in the copper content in the seawater and sediments surrounding the salmon farm.[302] Scientific studies in Norway have shown elevated concentrations of copper in sediments and in marine life around salmon farms and net washing stations.[303]

Scottish studies have found similar results. In 1995 the Highland River Purification Board (later SEPA), described how 4.5 kilograms of copper leached from a single salmon net in just three days.[304] A secret survey carried out by SEPA in 1996–97 found that sediments directly beneath cages and within 30 metres of the farms were "severely contaminated" by both copper and zinc at seven out of the ten farms surveyed. Copper concentrations were elevated by up to 25 times and zinc by up to six times. The report concluded: "It is likely that the high concentrations of metals, together with high levels of toxic substances such as sulphides and ammonia, will represent a significant barrier to the re-colonization of the benthic sediments at the affected stations in the various sea lochs visited."[305]

Scientists have also found zinc contamination under salmon farms in New Zealand,[306] and salmon farmers in Australia use copper-based paints to reduce the threat posed by predators. "The industry has in the past avoided the use of conventional anti-foulants on net cages, but has recently obtained a permit from the National Registration Authority to use copper-based anti-foulant nets in an attempt to combat seal attacks during frequent net changes of unprotected use," explains Darby Ross of the Department of Primary Industries, Water and Environment in Tasmania.[307] Fish farmers in Australia have more excuse than most, as seals are the least of their worries. Attacks by great white sharks are now commonplace throughout southern Australia (in northern Australia it is crocodiles).

It is difficult to feel sympathy for sea cage salmon farmers, though. If salmon farms were on land there would be no need for antifouling paints or many of the other toxic chemicals pumped into the sharks' swimming pool, predator attacks would not happen at all, and, perhaps more seriously, copper and zinc contamination of farmed salmon would not be a consumer health problem. The Australian government's "National Residue Survey Results for 2001–2002" detected copper and zinc contamination in 100 percent of farmed salmon tested (60 out of 60 samples).[308] By discharging hazardous chemicals into the sea and by causing contamination of farmed salmon, salmon farmers appear to pose inherently more danger than the great white shark any day of the week.

Malachite Green—A carcinogenic chemical contaminating one in seven farmed salmon tested in Europe in 2002

The only thing green about this chemical is its colour. Malachite green is a synthetic triphenylmethane dye that was spawned in the laboratory. It is one of the industry's longest serving chemical weapons, and its use in fisheries dates back to the time of Rachel Carson's *Silent Spring.*[309] Salmon and trout farmers have used malachite green since the 1970s[310] because it is so effective at killing the fungi and parasites that plague farmed salmon. However, now it threatens to blow Chilean and Scottish salmon farming out of the water.

Health agencies in the Netherlands, Spain and the UK have all refused entry to Chilean farmed salmon contaminated with malachite green, and during 2003 the European Commission's Health and Consumer Protection Directorate issued about a dozen "Rapid Food Alerts" warning consumers that Chilean farmed salmon could be contaminated with the chemical.[311] These are not isolated incidents but involve the entire industry, including some of the largest salmon companies in the world. In 2002, Nutreco, for example, was fined for the illegal use of malachite green.[312]

Scottish farmed salmon is so contaminated that ten to twenty percent tested positive between 2001, when monitoring first began, and 2003.[313] As a result, the European Commission has threatened to ban imports of farmed Scottish salmon to the European Union.[314] (The UK banned use of malachite green on fish farms in 2002.) Ireland, New Zealand, the Faroe Islands and Norway have all been guilty of using malachite green illegally. For example, malachite green has been widely used in Norwegian salmon farming since at least 1989, when 26 kilograms were consumed. Official figures show that its use peaked in 1991 at 114 kilograms, falling to 47 kilograms in 1995 and 27 kilograms in 2000.[315] Svanhild Vaskinn of the Norwegian State Food and Beverage Inspectorate told Intrafish in 2001 that "Malachite has not been assessed and is therefore prohibited from use on fish that will be consumed."[316]

Unfortunately that has not deterred some salmon farmers, who have shamelessly cut costs by using malachite green, which is twenty times cheaper than the less effective alternative. As Professor Ron Roberts, vice-chairman of the Animal Health and Welfare Committee of the new European Food Safety Authority, said in 2003: "There is a licensed product available for use in salmon egg hatcheries, called Pyceze. It works to some extent but is some twenty times as expensive as malachite and has no residual effect in protecting the eggs between usage, so has to be used daily, at full therapeutic dose level."[317] Bronopol (trade name Pyceze) is manufactured by Novartis and has been used in Norway since 1999. It was available for use in the UK in 2001.

Malachite green has long been suspected of causing genetic mutations that can lead to malignant tumours in humans. These cancer-causing properties were addressed in reports published in 1999 by the UK Department of Health's Committee on Mutagenicity and Committee on Toxicity.[318] However, salmon farmers, the chemical industry and the government must have been well aware of the toxic and carcinogenic effects of malachite green since the 1960s.[319]

In January 2001, the US National Toxicology Program issued the following warning about malachite green: "Because of its effectiveness this chemical is considered to have a high probability of abuse . . . the use of this product could result in significant worker exposure and the effluent from the aquaculture facility could enter the water supply resulting in exposure of the general public through recreational activities and drinking water. Finally, the use of malachite green in food fish could result in human consumption of malachite green residues."[320]

In January 2003 a European Commission science panel recommended classifying malachite green as a toxin that poses a risk of birth defects and harm to public health.[321] It has been reported to be injurious to the human eye, it caused skin problems in six of eleven eczema patients,[322] and in 2004 the US Food and Drug Administration received information from the US Fish and Wildlife Service and the Centers for Disease Control and Prevention about a possible correlation between hatchery workers' exposure to malachite green and the formation of acoustic neuromas (tumours that can lead to hearing loss, imbalance and brain stem compression).[323]

Two recent initiatives on either side of the Atlantic could be the final nails in the coffin for malachite green use in fish farming. A meeting of the US Department of Health and Human Services National Toxicology Program in February 2004 brought official confirmation of malachite green as a carcinogen a step closer. Malachite green was "nominated for toxicity and carcinogenicity studies due to the potential for consumer exposure through the consumption of treated fish."[324] And in October 2003 a meeting of the World Trade Organization's Committee on Sanitary and Phytosanitary Measures agreed to the European Commission proposal to set minimum required performance limits (MRPLs) for malachite green in farmed fish. MRPLs for malachite green were formally adopted in November 2003 and are due to come into force in December 2004.[325]

Whether the new measures on malachite green are successful or not is somewhat irrelevant. The UK's Veterinary Medicines Directorate has said that "although the use of malachite green was banned in the UK [in 2002], estimates indicate that we can expect to see residues up to around June 2006, and possibly for longer."[326] Malachite green, like teflubenzuron, ivermectin and emamectin benzoate, is so persistent that even if salmon farmers clean up their act right now, its "lingering legacy" will still be found in food for a long time to come.[327]

Shutting the Cage Door after the Salmon Have Bolted

It is clear from these chemical case studies that salmon farms are slipping through the net. In a complete reversal of the precautionary principle, the policy of governments has been to issue licences for toxic chemicals first and ask questions second (and then only if they have to).[328] So instead of completing a risk assessment before a chemical is approved for use, we have the unsatisfactory situation of assessments taking place after the event, if they occur at all, far too late to ensure either environmental or human safety.[329]

The UK government's Post-Authorisation Assessment Programme illustrates all that is wrong with the current chemicals licensing system. This £4 million pesticide probe was first mooted in 1994, but the salmon farming industry successfully delayed it until 1999. Since then it has suffered a series of setbacks and faces an uncooperative attitude from the industry. The five-year study has taken so long to publish its findings that many of the chemicals being assessed will have been phased out in favour of a new generation of chemicals.[330]

Plugging the research gap in the environmental assessment of chemicals is of primary importance. A report published in 2002 by the Scottish government concluded that a great deal of research needed to be carried out "on the toxicity of emamectin benzoate, teflubenzuron, copper and zinc to benthic organisms commonly found in Scottish sea lochs; more information is required on the long-term effects of cypermethrin, emamectin benzoate, copper and zinc on sediment associated organisms; more information is required on the dispersion, fate, and potential long-term effects of multiple cypermethrin treatments (at single and multiple farm sites) within a loch system; more information is required on the potential effects of concurrent emamectin benzoate treatments at several farm sites within a loch system."[331]

For chemical companies, less is more. The less public scrutiny, the more chemicals will be sold and the more profit. It is clear that there is a great deal of money at stake here. The worldwide parasiticide market for terrestrial livestock was worth US$3 billion in 2000,[332] and companies such as Novartis and Schering Plough apparently see oceans of opportunity in the sea cage fish farming sector. The market for chemical products for sea louse control currently accounts for less than one percent of global parasiticide sales, but it is an emerging one.[333]

Sea lice infestations can reduce the yearly market value of farmed salmon by up to twenty percent due to cosmetic effects, poor growth and fish mortalities.[334] In 2002 the Scottish Salmon Growers Association estimated the costs of stress on infected fish and loss of growth due to sea lice infestation alone cost the Scottish salmon farming industry £13 million per annum. Big bucks are involved: "Discussions with pharmaceutical companies reveal that there is a market of £4-£5 million for medicines, and when this is added to the costs of administering medicines including hardware and labour, the costs of accidental treatment mortalities during bath administration, and the costs of down-graded product at harvest, a total cost per annum of £20-£30 million is acceptable to most in the industry."[335] If you extrapolate that figure around the world and take into account inflation, you have a conservative estimate of £200 to £300 million per year to be spent on parasiticides. Acceptable costs to the salmon farming industry perhaps, but the marine environment, shellfish, wild salmon and consumer health bear the brunt of these savings—a heavy price to pay for cheaper salmon.

A War with No Winners, a War with No End

In this warped chemical weapons race there is a recurring nightmare: Government authorities license a chemical knowing full well it is toxic, protect the chemical company from public scrutiny, and when a risk assessment is finally published years later (after the target's resistance to the chemical has made its use redundant anyway), a new chemical takes its place.

Salmon farmers are fighting a losing battle against their nemesis, the sea louse. As Craig Orr of Watershed Watch says: "Lice rapidly develop resistance to all chemical therapeutants (three to five years) and, as long as we practice open-net-cage aquaculture, we'll always need newer, better and more expensive drug and lice treatments."[336] So concerned are salmon farming nations that an international European Union-funded project named SEARCH (SEAlice Resistance to CHemotherapeutants) involving Norway, Scotland, Ireland and Canada has been set up to combat the problem of sea lice resistance.[337]

In the absence of new treatments, salmon farmers use existing chemicals in greater quantities[338] or in combination. "Integrated Sea Lice Management"—a phrase often used by the salmon farming industry—merely means using several different chemicals instead of just one.[339] This is done without taking into account their synergistic effects—the so-called cocktail effect.

The chemical industry continues to put new and dangerous chemicals on the market and is attempting to "harmonize" the use of chemicals worldwide. This would mean that chemicals available in one country are also available in others.[340] The Veterinary International Co-operation on Harmonisation, launched in 1996, is "aimed at consolidating technical requirements for veterinary product registration."[341] A September 1997 roundtable discussion in Edinburgh on "Progress with Registration of Drugs and Vaccines for Aquaculture" sought "world-wide co-operation to gain approvals of drugs." Similar initiatives have been developed around the world. Salmon Health, for example, was developed in Canada "to assist pharmaceutical companies to compile data submission dossiers to meet the requirements of regulatory and licensing agencies." It is run by the Canadian Aquaculture Industry Alliance and is funded by the aquaculture and manufacturing sectors (pharmaceutical and feed) and by government agencies. Over five years the number of approved therapeutants for salmon farmers in Canada increased from three to six, with temporary registration of three additional compounds.[342] A cynic might see this initiative as nothing more than an attempt by the chemical lobby to short-circuit the chemicals registration process and fast-track chemicals globally.

There is an eerily familiar sense of déjà vu as the same mistakes are repeated over and over again: the name of the chemical may be different but the problem remains the same be it canthaxanthin, dichlorvos, azamethiphos, cypermethrin, teflubenzuron, ivermectin, emamectin benzoate, TBT or malachite green.

Closing the Net

To avoid a "Silent Spring" of the sea we must curb chemical use in the entire sea cage fish farming sector now (sea bass, bream, barramundi, kingfish and tuna farmers are already using similar chemicals). Even then the lethal legacy of sea cage salmon farming will be with us and our children (if we don't become too impotent to have them) for a long time to come. Over 40 years after *Silent Spring* was published, the chemical Carson exposed, DDT (widely banned in the early 1970s), is still being found in farmed salmon along with PCBs, dioxins and other contaminants. It is a dreadful prospect to think what we will be finding in 2044.

The salmon farming industry must tackle the causes, not the symptoms of addiction. For the long-term health of the marine environment and consumers, the industry must stop discharging contaminated wastes directly into the sea and start ripping out sea cage salmon farms. If salmon farmers adopted closed containment technology to treat their chemical wastes, environmental impacts would be reduced at a stroke, yet closed containment systems are dismissed as too costly. A SEPA report in 1998 concluded: "Capital costs are likely to be prohibitively expensive for all but the largest producers. Although land-based systems currently offer the

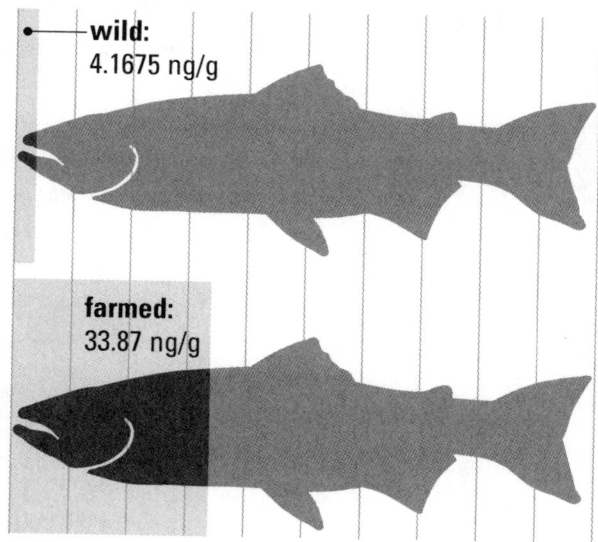

greatest potential for containment and treatment of wastes following chemotherapeutant use, the systems are not viable for commercial salmon production under present economic conditions."[343] Similar studies on wastewater treatments and closed containment technology have been conducted in British Columbia but, all too predictably, are considered "uneconomic."[344]

Even for in-feed treatments there is a solution other than dilution. "What we're developing is a carrier for medicines that will allow medicine to be added to the fish food and then come out in the stomach of the fish," says Lynne Wallace of Ensolv Ltd. in Scotland. "From an environmental point of view, this will reduce the amount of waste going into the water column. What'll happen is that if the product goes into the environment, the therapeutant chemical won't leak into the water, and it'll be able to be collected, as opposed to having the waste discharged into the water. We need to get fish farmers on board. If they could reduce the amount of environmental waste, that would improve their position with respect to licensing, and how much of a product they could use. At the same time, we need the pharmaceutical companies because obviously it's their medicines that would go into the carrier."[345]

It is far too late for the hundreds of thousands of tonnes of chemical wastes dumped into the sea by salmon farmers since the 1970s, but it is at least a step in the right direction. Other sensible solutions include reducing stocking densities, increasing fallowing periods or using biological controls such as cleaner fish, which work in harmony with nature. As Rachel Carson said on CBS television shortly before her death in 1964 from breast cancer: "Man is a part of nature, and his war against nature is inevitably a war against himself."[346]

Dying of Salmon Farming

Alexandra Morton

To Broughton, whom I love.

The first fish farm I laid eyes on was passing the windows of my home in 1987. It was an insignificant steel structure creeping obediently behind a black and pink tugboat. I'm always glancing out my window for whales, and I raised my heavy 20 x 80 binoculars for a better look. The profile, a cross-hatching of steel uprights and buoyed walkways, was unique. This must be *a fish farm.*

The newcomer looked so benign, slow and low against the spectacular vista. The thought "Good idea" passed through my mind.

At that time, Echo Bay was a micro-community of only about 100 people, continually threatened with the loss of our school, loss of the post office and loss of the community itself. We are so fragile a group that if one family leaves, we could float away, as all our once neighbouring communities have done. So I hoped this new industry might attract new families, offer local employment and shelter me from sudden storms as I wandered the inlets studying whales. What I failed to grasp was that *it* was the storm. I wish now we had done everything possible to stop that first farm. But we had no idea.

I settled in Echo Bay in 1984. My late husband, filmmaker Robin Morton, and I prowled this coast for a home like a pair of nesting geese. We surveyed bays, ticking off their attributes against our short but immutable list: protected waters, whales year-round, post office, and children for our infant son to grow with. In October of that year we followed matriarch orca Scimitar and her family east into Fife Sound, portal to a world where land meets water in a million places. Richly scented estuaries smelling of mud and wild mint, sheer cliffs adorned by clinging cormorants, and tiny islets crowned with bonsai

Fish farm under tow.

cedar bent in homage to the prevailing winds were threaded like jewels by the swiftly flowing tides. That first day I fell in love and adopted this wilderness as a home where I could raise my children and study whales.

All the Wrong Places (1988)

When a scientist makes a wild place her home and study site, organized data about that ecosystem begins to trickle out into reach of the greater human consciousness. I began publishing research papers on the secretive lives of the transient orca, on the explosive return of Pacific white-sided dolphin to this coast and, with co-authors, on the diet of killer whales. Every morning arrived ripe with the possibility of learning something new. The fundamental act of science is learning how to see connections, but no wilderness lays its secrets bare. For me, a person who did not grow up in her study area, guides are essential for explaining the difference between human perception and the reality of the natural world. The killer whales I came to study are salmon predators that could indirectly answer my questions, but fishermen are also salmon predators, and they could talk.

Wind and tide are easy to see, but what do they mean? Why is this chinook so short and deep-bodied? Why is the water brown today? Why do orca males travel with the tip of their fins exposed only in glacial melt-water? How many salmon should be here now? What route did they travel? When will this storm end?

The fishermen of my community had answers to these questions and many others. Casual encounters on the post office dock often led to long conversations and copious scribbled notes. So I have no excuse for carelessly shrugging off their early grumblings about where the salmon farms were dropping their anchors, how many fish were put in the flimsy netpens and what kind of fish they were farming. Perhaps it is because I am from New England, where the old-timers distrust anything shiny or new. But these fishermen persisted. "They are putting them things in all the wrong places."

In the winter of 1988 a flashy Boston whaler came zipping into Echo Bay. Everyone noticed. It was the wrong type of boat for winter. Locals run beat-up fibreglass boats, whale watchers use inflatables, fish farmers come and go in aluminum boats, but this looked like a tourist vessel, as out of place as gumboots in New York City.

"Hi, we're from the provincial government. I'm Alex." A young man with strawberry blonde hair and the same name as mine extended his hand. "We're here to find out where you guys don't want salmon farms," he told those of us waiting for the mail plane.

This sounded positive, and I volunteered myself as a guide to show them what I knew about whale movements, where kayakers camped and seals hauled out. I pounded on my neighbours' doors. "Hey guys, here is your chance to get the farms out of those *wrong* places." But no one seemed as impressed as I was. The fishermen were not eager to share a lifetime of secrets—what they knew about where the wild fish schooled, travelled, held and fed—and I could not understand. They trusted me with this information but not the government—why?

In the end the top fishermen of Echo Bay piled into Bill Proctor's well-seasoned little *Twilight Rock* to meet with federal and provincial people sent to pinpoint the areas of Broughton Archipelago with the greatest biophysical capability. There was Bill, born and raised here, with over 40 years of highlining wild salmon coastwide; Chris Bennett, who could outfish the orca; Al Munro, a man who could find prawns when there were none; and me, a happy fool. We were off to save our home, but the Broughton threw a lot of water at us that night. The howling gale almost seemed a warning—Don't do this to me.

Rough wool toques bob among white starched bureaucratic collars in my pictures of that night in Alert Bay. The vital organs of the Broughton were circled and notated on chart after chart. Upwellings here; Knight Inlet chinook salmon school there; pink salmon wait out the ebb-tide here; largest, most abundant prawns here, young salmon rear here . . . The Broughton was safe and salmon farming was limited to non-essential areas. Right?

Wrong.

In a few months we pored over the chart produced by this government team. The Broughton was painted red, yellow and green. Green areas were available to salmon farms, yellow meant "proceed with caution" and red was defined as areas where "no application for fin-fish farming would be accepted." They couldn't even apply for the red zones, which marked exactly the areas most important to the wild stocks. The map was not perfect but it was workable. Then, in a betrayal of public trust that set the tone for coming years, farm after farm appeared in the "red zones" until there were more farms where they were *not* allowed than where they were permitted.

"Who is in charge of this mistake?" I asked again and again, and the circle game began. The Ministry of Agriculture and Fisheries said, "Not us, we only issue the licence to operate; the Ministry of Environment [MoE] is in charge of where the farms go." But MoE said, "Oh, we send out referrals to the Department of Fisheries and Oceans [DFO] and they tell us whether the farms will impact wild fish." The DFO field officers said, "We *did* say no to those sites, but management wouldn't listen to us and some of us quit."

Salmon farms were an unruly orphan no one would claim. Guidelines were issued to give the appearance of management, but the devil lies in the details. "Guidelines" are not regulations; they are suggestions. Loggers also use "guidelines" to log in fish habitat.

Bacterial Kidney Disease in Viner River Chum (1989)

The first year of mysterious wild fish disappearances in the Broughton Archipelago was 1989. Chum or "dog" salmon, have a characteristic way of leaping. They are a large salmon—many are over nine kilograms—and they slide out of the water, punctuating a half circle with one leap after the next. For this reason we always knew when the chum salmon were coming home to spawn in the Viner River. As the tide floods, these fish explode at the surface, ten or twenty in the air at any one moment. They also school and roll at the surface like porpoises. Every fall orcas of the G clan would come into the Broughton specifically to fish this stock. In the big years, seine boats came too.

In 1989 there were so many chum visible at the surface that my neighbours who were fishermen and DFO patrolmen mused this would be another big year for the Viner River, 70,000 chums or more. But the fish never made it upriver. Instead their large tiger-striped carcasses rotted at the mouth of the river, their eggs imprisoned in bodies that for some reason could not go the last 500 metres and died too soon for the next generation to be born. At the time I thought this merely a quirk of salmon biology. I had eyes only for whales and never thought to find out why all those fish had died.

It was a few years later, when I began researching the literature on salmon farms, that I learned some farms in the Broughton had big problems with bacterial kidney disease (BKD) in 1989. I found a paper reporting that when chum salmon were exposed to BKD, 100 percent *died.*[1] Is this what happened to the Viner chum salmon? Today, fourteen years later, biologists are still trying to figure out what happened to them, because elsewhere on the coast, chums are coming on strong.

Furunculosis in the Scott Creek Hatchery (1991)

When Scott Cove Creek was logged in 1918, a dam was built to raise the level of Loose Lake on Gilford Island so the logs could be floated to the creek. At intervals the dam was opened and the logs sluiced to the sea. When the logger was done he walked away, leaving the dam in place and generation after generation of coho to beat themselves senseless against it. A group of Echo Bay residents including loggers, fishermen and DFO patrolmen and an American tourist took it upon themselves to bring these fish back. The dam was removed and the coho coaxed back into existence using local stocks and a hatchery. For ten years every man, woman and child in Echo Bay participated in this miracle by running back-breaking relay races down the hillside with seven kilograms of live fish in a bucket of water or by supplying food for the runners. The success was unmistakable. Meticulous care resulted in the loss of less than three percent of the adult fish brought into the hatchery and the release of tens of thousands of baby fry each spring.

In 1991 the usual team of volunteers was running the hatchery once again when every morning brought the shock of big, beautiful fish lying dead at the bottom of the troughs. Each was branded with red-ringed, angry boils erupting liquid and infection. DFO diagnosed furunculosis. "Never heard of it," quipped hatchery president Bill Proctor, "but the drug they gave us worked a miracle."

We wrestled every fish and injected it with oxytetracycline. Over a quarter of the fish died before we could treat them, so we were happy to save most of them. However, we all knew that without the antibiotic a lot of wild fish must be dying. It was natural for us to wonder where this virulent bacteria had come from.

There are no secrets in Echo Bay, too few people to isolate truths, and so in Bill's dark, cedar-panelled living room, we learned from a fish farm manager that one of the other operators in the area had furunculosis break out in its hatchery. The fish farm representative said this operator went ahead and put those Atlantic smolts into netpen farms in the Broughton. There the infection had raged, and the operator attempted to quarantine the fish by keeping boats away. Helicopters moved drugs, personnel and messages because the operator was afraid using the radio would alert locals to the epidemic. The fish farm representative's firm had felt threatened by this epidemic and arranged to buy the other operator's farms in the Broughton.

"What did you do with the fish?" we asked almost in unison.

He explained they killed the farm fish that remained in the hatchery.

"But what about the fish already in the Broughton?" We hung on his answer.

He said they treated them with antibiotics and left them in the netpens. Any survivors they would sell.

This was a lot to process, but I had to ask, "Were the fish you treated completely cured, or did they remain infectious to the wild fish?"

Our informant told us the farm fish remained infectious. We as a community, united in restoring wild fish, reeled and railed. How could DFO allow this?

"Where did the bacteria come from?" I asked. Silence . . . Then a reply that became an industry standard: "From wild fish."

This roused my dormant antennae. Were we really supposed to blindly accept that an epidemic of the same disease had occurred simultaneously in farmed and wild salmon? Where was the supporting evidence for this hypothesis? I began sniffing about. Where did the fish come from? A hatchery in Scotland. Was there any history of furunculosis there? Yes, salmon from that same hatchery spread a virulent strain of furunculosis to 74 wild rivers in Norway, killing thousands of farmed *and* wild fish.[2] I knew what strain of furunculosis had killed our coho, so I wrote to scientists in Norway to ask what strain came from the hatchery in question. Armed with that I asked the local farmers what strain their fish had . . . silence. I asked DFO . . . silence. I went to meetings with my question . . . angry, explosive silence. I lost trust in the system; I felt it was working to hide the truth.

Furunculosis and Kingcome Inlet Salmon (1993)

In 1993 a young farmer at the gas dock complained that they had a "triple dose of furunculosis." Feeling more like a bloodhound than a whale researcher, I began to circle. It seemed that Scanmar Ltd. Atlantics had been infected by a strain of furunculosis that was resistant to all three antibiotics approved for use on salmon farms in BC (Ron Ginetz, DFO, pers. comm.). The provincial government called it an "emergency" (Al Castledine, MAFF, pers. comm.) when the epidemic spread ten kilometres in a few days to infect BC Packers' farms.[3]

A pattern began to appear. As in the previous epidemic, one company bought the other (in this case, BC Packers bought the Scanmar sites) and left the diseased fish in the water. Surprisingly, DFO not only allowed this, but also okayed the use of erythromycin, normally banned for use in fish destined for human consumption because it is an important drug to combat human lung infections.[4]

To get from Scanmar to the BC Packers farms, the pathogen crossed the migration route used by all the wild Kingcome Inlet salmon, as well as an area closed to commercial fishing since 1968 when it was recognized by DFO as an important chinook salmon rearing area. I begged DFO to test the wild fish. The high drug resistance of this bacterial strain would have identified it as brilliantly as a neon sign blinking "farm origin, farm origin." DFO tested one pink salmon from the Kakweiken River, 55 kilometres away, found the bacteria and claimed to have lost the culture before being able to assess the drug resistance. I offered to collect more fish for them at no cost, but they declined. They did not want to look at any more wild fish from this area.

Once again I began searching the public record only to discover that Scanmar had been looking for a source of Atlantic salmon eggs in the western United States where there was a problem with "atypical and drug-resistant strains [of furunculosis] . . . not known to occur in BC."[5] I could not find out if Scanmar actually made the purchase. As if none of this was quite enough, BC Packers had an accident at one of the newly acquired farms in Wells Passage, and tens of thousands of diseased farm salmon were released just as the Fraser River sockeye were on their way south through Queen Charlotte Strait to spawn.

The next spring sport fishermen in Kingcome Inlet experienced a serious decline of the chinook stocks. It affected the vigorous "feeders" following the oolichans into the head of the inlet. It affected the 18- to 23-kilogram spawners destined for the Kingcome and Wakeman rivers. It even affected the juvenile "shakers." When logging damages a stock of fish you lose each successive younger generation. When overfishing damages a stock you lose the older fish. But disease takes all generations simultaneously, and this was the signature of this collapse. Three fishing lodges moved out of the area and another closed. But the most alarming concern was the fact that chinook salmon with tags from as far away as the Elwha River in Washington state have been caught in Kingcome Inlet, which meant fish that did not die outright may have become carriers to rivers throughout the eastern Pacific.

Through all of these epidemics DFO and salmon farmers repeated their mantra: "No evidence has been found of disease transfer from farm to wild salmon." By now I knew that one reason no evidence had been found was that no one was looking. I set my killer whale research aside and spent three months reading all the scientific literature about salmon farms I could lay my hands on. I wrote 10,000 pages of letters to government, salmon farmers and scientists. I was convinced for a time that if I could simply line the words up correctly, I could inspire action to prevent any further loss of wild salmon.

But the answers from bureaucracy and industry were evasive; they reminded me of a horse I rode as a child. No matter how hard I tried, how properly I followed protocol, how much pressure I applied, I could not make that horse head straight for the jump and go over it. Similarly, the answers I got darted off in any direction but did not answer my questions. What strain of bacteria infected the farm fish? How many seals are being shot in the Broughton? Why is our water turning red and causing burning sensations on our skin? Always the reply came back, "Dear Ms Morton, thank you for your concerns . . . " And on the letters went, telling me all was taken care of and I could run along and play. But after a life of studying wild animals I recognize evasive behaviour, and what I read in their letters was: *We don't want you looking into this anymore.* That really caught my attention.

AHDs Drive Whales out of the Broughton (1993)

In 1993 the relationship between fish farmers and the federal government became much clearer to me. A French film crew was with me as I searched for orca through the inlets that January, and they recorded my shock as I ripped the headset off my head and rubbed my ears. Used to the silence of the underwater world, I often turn my tape recorder on with the headset in place, but on this soft grey morning the silence was gone. In its place a mechanical cricket-like chirp screamed and would continue screaming for the next five years. I later learned it was 198 decibels, the same volume as a jet engine at takeoff, and was designed to hurt the ears of seals and keep them from attacking the high-calorie domestic fish protected only by nets.

Four farms started using these "acoustic harassment devices" (AHDs), and over the next months I watched in horror as one killer whale family after another encountered the wall of sound and fled, never to return. Whales "see" with their ears, keeping in touch with family members and deciphering their precise echolocation clicks to find food. Hearing is their primary sense. These sounds, with their capacity to damage marine ears, were akin to a room full of needles zinging toward the eyeballs of animals like us, which depend on vision. You or I would be unlikely to re-enter that room for a long time, even if it was integral to 10,000 years of our ancestral history, and it was the same for the whales.

Section 78 of the Fisheries Act states very clearly that anyone who disturbs whales faces a fine of up to $500,000 and two years in jail. The whales' response was unequivocal, yet the DFO, whose duty it was to uphold this act, chose to turn a blind eye. At first officials pooh-poohed my concerns. Then they did a study using harbour porpoise instead of whales. When the AHDs were turned on, the harbour porpoise left. But instead of publishing their research and upholding the Fisheries Act, they shelved the study and allowed the farmers to continue. I upped the ante and published my findings as a peer-reviewed scientific paper,[6] and still the DFO would not act. Finally, five years later, the fish farmers turned off the AHDs voluntarily because they didn't work. The seals were attracted to the devices in what is dubbed a "dinner bell effect," willing to risk their hearing for the high-calorie farm salmon. However, the damage had been done and even today, ten years later, the resident whales still distrust this area and most with youngsters will not come back.

My head was spinning. Why risk the abundant, highly renewable wild salmon for a flash-in-the-pan industry plagued by disease? Why allow salmon farmers to displace the whales government went to great expense to protect just a few nautical miles away? Why consult the locals, promise compromise, then break those promises? It had taken the Broughton 10,000 years since the glaciers last receded to reach this current state of abundance. Was it possible I would record its demise? Could I possibly be a whale researcher living where there were no whales? Should I leave or stay?

Bureaucrats and Allies (1995)

I punched out letter after letter, connecting the dots in the plainest language I could summon. The muscles of my face twitched with rage I knew could not be released. Dammit, "they" simply were not going to kill something I had grown to love. Love is an incorruptible force that only grows stronger when denied or threatened. I began pacing my world like a caged animal.

"Who in god's name is running this place?" I asked a neighbour prawn fisherman, "because the politicians have no idea what is going on here." The environment minister had just tried to tell me "there are no salmon farms on wild salmon migration routes," when in fact it was physically impossible for a wild salmon to reach a river in the Broughton without passing several salmon farms.

Albert smiled into his flowing beard and leaned forward. "Bureaucrats, Alex. They are the guys running this show."

I began to notice this layer of government, never elected, nameless on public documents, which never moves as governments rise and fall. Puppet politicians strut and preen to distract us while bureaucrats quietly pass memorandums of "understanding" from Ottawa to Victoria, immune to a lack of votes from afflicted coastal communities. They construct our reality. They slide back and forth between employment with government and industry, incestuous at best.

Only a few weeks earlier I had sat across the desk from a bureaucrat to request permission to tie my floathouse to the coast of British Columbia. Crisp in his well-tailored suit, the bureaucrat smiled brightly and said, "Not possible."

"Excuse me? We have a school, a post office and a community but no land, and now we can't live in our floathouses?" That effectively spelled the end of Echo Bay.

"Well, yes," he explained implacably, a *memorandum of understanding* had been signed in Ottawa permitting foreshore leases exclusively to fish farmers, loggers and floating resorts. A way of life on this coast, over 100 years old, was terminated by the stroke of a pen, and who knew, who was consulted? No one I knew.

I took this news home to my community. In our cozy floathouses we pondered this over sumptuous potluck dinners of prawns, salmon, venison and clams. We were as doomed as our dinners. The best analysis was that we were unwanted witnesses to the corporate interests who had decided they wanted to use the Broughton. We had to be cleared away in a tried and true method used over the ages to take the commons from the common folk.

"How do we stop this?"

"Allies" was all my fisherman friend and neighbour Chris could say. "I know you gotta have allies."

In 1995 the provincial government selected representatives of the salmon farming industry, government, fishermen, environmentalists and coastal community residents to undertake a salmon aquaculture review. Laurie MacBride, managing director of the Georgia Strait Alliance, said, "This is all because of your letters, Alex." I was relieved. Now we could present the impact of this industry on wild fish, whales, communities and coastal health in general and bring some sense to the helter-skelter destruction, threats and conflict—and I could go back to studying whales.

The review was a gruelling process to which I was not invited, so I had to foot my own bills. However, my neighbour Bill Proctor, long-time fisherman and resident of Echo Bay, was invited and we teamed up. All winter he carried me and my infant daughter Clio by sea, and I took us the rest of the way by car to the monthly three-day meetings in Tofino, Campbell River, Port Hardy and Nanaimo.

An open microphone inspired resident after resident, voices quaking with stage fright and passion, to rage, rationalize and plead against expansion of salmon farming. On the flip side, fish farm employees and support industries got up and stated opposite opinions. I wrote three reports, on whales, disease and impact on my community.

However, it became clear the terms of reference blocked serious input from the gathered crowd. The review committee would not consider research on aquaculture impact from outside BC, would not consider "anecdotal" local evidence and would not fund the studies required to produce made-in-BC science. It was a box canyon and we didn't have a chance. With no supporting evidence and without consulting Bill Proctor, president of Scott Cove Hatchery, whom they had invited to their table, they decided the 1991 and 1993 epidemics in his hatchery must have been due to "a change in handling procedure." Only they could make statements based on pure speculation.

What had glistened so promisingly crumbled to rust as it dawned on me, "This is a rigged game, and no truths will come to light here."

But there was a silver lining. It was unintended, but as handlers pushed us further from resolution, a miracle blossomed. In the crucible of corporate and government alliance I met David Lane, true statesman for wild salmon with the T. Buck Suzuki Foundation; Laurie MacBride of Georgia Strait Alliance, who could actually speak bureaucratese; the glorious Lynn Hunter, ex-member of Parliament; leonine Catherine Stewart of Greenpeace; the indomitable

First Nations wise woman, Yvon Gesinghaus; tenacious lawyer Karen Wristen; and silverback male, Jim Fulton from the David Suzuki Foundation. Like any assemblage of earth's matter under pressure, we aligned, we became *allies*. We are now the Coastal Alliance for Aquaculture Reform (CAAR), with many members.

The Pacific's First Atlantic Opening (2000)

On a warm, still morning, August 3, 2000, an angry fisherman's voice blasted from the marine radio mounted in my speedboat: "Got another damn Atlantic, think that makes 48 for the day, my net's full of more f—g farm fish than sockeye!"

Like whales calling back and forth across Johnstone Strait, one fisherman after the next got on the radio and reported similar numbers. I flicked the marine radio to the weather channel. "Low clouds and fog, windy 25 to 30 knots from the west."

Too windy and foggy for me, so I called Dennis Richard's water-taxi. Dennis picked me up and we headed into the strait. The first skipper scrunched his face at my request. "You want WHAT?" he bellowed over the slap of waves and the motors of both boats idling. "*Atlantics?* What are you—DFO?"

The next boat thought I was a fish farmer and threw a farm fish *at* me. The rage was palpable. This was going to be more difficult than I thought. I got on the radio and at each channel asked any vessels with Atlantic salmon on board to get back to me, "please" . . . silence.

"If you want these fish to be counted, please let me know how many you have caught," I tried again, expecting a retaliatory streak of profanity to get off the radio. I deckhanded on a fish boat for four years and knew no fisherman likes to be called out of the blue and have his numbers demanded! But I was at a loss.

Then Dennis grabbed the mic out of my hand and yelled into it, "You wanna stop these farms, call the lady here with the pen. If you like the farms, stay quiet." Yikes.

And the dam opened. "*Blackfish Sound* [that's me], *Dream Weaver* here. I got 37, all big buggers, off Robson Bight." "*Outsider* has 30." "Put *Winning Edge* down for 60, 30 in the last set." "*Valerie* here, I have some on deck for you."

I was writing at top speed. We pulled alongside as many boats as we could in the building breeze. The aluminum water-taxi bobbed violently while the larger gillnetters provided some lee from the wind, and the Atlantics started flowing over the gunwales. They were identical—"looks like they been through the same cookie cutter," remarked one fisherman. I counted 1,638 Atlantic salmon caught that day in upper Johnstone Strait.

I measured, weighed, sexed and opened every one I could lay hands on. I wasn't sure what essential data to collect, but *I* wanted to know what they had been eating, so I recorded stomach contents as well.

An embarrassed salmon farming industry belatedly admitted an escape. An underwater screen had not been put in place as a packer carried a load of live farm salmon to the processing plant via Johnstone Strait. While wild salmon are killed and generally cleaned at sea, the flesh of farm salmon is so soft they have to be delivered alive to the plant. The packer crew admitted to losing 4,500 fish from the hold two days earlier. The fraction of the fleet I had spoken with had caught almost half the escapees 48 hours later.

The next time the fishermen put their nets in the water it was worse. This opening was in steep-sided Tribune Channel, where I had often watched whales sleep. The high boat got 250 Atlantic salmon in a day, and 40 to 50 was the norm. They caught more Atlantic salmon than pink salmon. "Can't believe these big fish will gill in broad daylight," quipped several. Once again all these fish looked alike, but different from the Johnstone Strait Atlantics, and the fishermen noted several had commercial pellets still caught in their throats. This could only mean they had been near a salmon farm, likely *in* a salmon farm, only minutes before capture.

All eyes turned toward the nearby Stolt sea farm in Sargeaunt Pass. Bill Proctor gave the manager a call. "Nope, we are not losing any fish," he replied. "Thanks though." They didn't know they were losing fish, but when the fishermen reported more fish with pellets in their throats the next morning, it became obvious fish had not only gotten out of a salmon farm, but they were *still* getting out.

Choking back early bravado, the salmon farmers admitted there had been a hole for an unknown time through which an unknown number of farm salmon had escaped from Sargeaunt Pass. The fishery closed and then reopened in a week. The hole had been sealed, but escaped fish were still hanging around and the fishermen hauled them aboard by the hundreds. The question of what to do with these fish arose. Had they been through all drug withdrawal times? Were they safe to eat? Should they be thrown overboard or was that release of contaminated waste? Many fish were not counted because these Atlantics had few to no spots, and fishermen were concerned they might be coho. There was a no-retention ban on coho, so many slipped back into the inlet waters, uncounted.

The fish farmers quickly said they would buy these fish and pay well. They offered fishermen coho prices for them, and that made it difficult for me to continue collecting samples. So I went to the wild salmon processing plant in Sointula where most of these boats were headed.

"You want to do what?" asked an incredulous plant manager.

"I would like to look through all the guts taken out of the Atlantic salmon, please." By now I was used to being considered odd.

He gave me a chair and a hairnet and instructed the forklift operator to place each cubic metre tote beside me. What a wealth of information lay in those heaps of intestines and hearts, and I didn't have to go chasing off after them. A professionally cleaned fish has all its organs removed intact in one clump. I could read the age and sex of the fish from the condition of the gonads. The crispness of the spleen's edges reported some measure of health. The stomachs gave up the fish's last meal, and the adhesion of one organ to the other revealed whether that fish had been vaccinated or not. I found a babysitter for my daughter and went to work. DFO and industry had told us escaped farm salmon were too domesticated to eat wild food; here was the proof one way or the other.

None of the fish caught in Tribune Channel during the first week of the escape had a trace of wild food in them. Some had pellets in their stomachs. One week later, two percent had attempted to feed. While some were experimenting with alder catkins, chips of wood and Styrofoam, four out of 497 fish examined had fed successfully, capturing both herring and young salmon.

When the fishery reopened one week later, I was turned away from the processing plant. In an e-mail labelled "dead fish talk" I had made the mistake of reporting on the wild food I had found in their stomachs. I would get used to doors slamming shut when a scientist looks at the wrong thing. An important avalanche of data was slipping away. I really wanted to know what the fish had found to eat in their third week of freedom. Finding wild food is the most crucial test any animal invading a new ecosystem must pass.

One fisherman after the next declined me access to his now valuable catch. I did not blame or press them; coho prices were huge compared to the extremely low price of the pink salmon, their target species. Then Calvin Siider called: "You can have my whole Atlantic catch, Alex."

Not giving him time to change his mind, I darted off to rendezvous with his vessel in Knight Inlet. I felt bad for his young deckhand, losing this part of his wages, but the hard set of Calvin's jaw made it clear he wanted his fish to go to science, and the fish were slurped into my boat. Calvin's fish completed the story. Upon release, none of the farm salmon I looked at were eating wild fish. Seven days later two percent were eating something including fish, and after 21 days of freedom, fourteen percent had passed the test with shrimp, herring, sticklebacks and other unidentified animal life in their stomachs—they could survive. Stolt figured it had lost 33,000 fish, which meant 4,620 could potentially be on their way to colonizing this coast.

Furthermore, by the second opening—officially called a "pink" opening, but dubbed the first ever Atlantic salmon opening in the Pacific—several fishermen tied one end of their net to the Sargeaunt Pass fish farm and made an interesting catch. While the vast majority of fish were swollen-looking creatures with enormous fat bodies accumulated around a small head, a few fish were fusiform, and sleek. The masses had soft greasy flesh that could be scooped and balled like mashed potatoes; the others had firm, muscular flesh. While the masses had few to no Atlantic salmon spots, these others were more typically coloured with dark black spots over their heads and bodies. And they were almost sexually mature while the others were years away from spawning. The males had "kypes," curved lower jaws, and their gonads were heavy with sperm while the females were ripe with large eggs.

It was my impression these fish came from other farms elsewhere on this coast, had found this farm and were hanging around the outside of the nets, making a living by eating pellets that drifted through the mesh. These fish were almost ready to find a river, and there were several good rivers nearby. These were the escapees most likely to spawn, and they set the stage for species invasion.

DFO officials said farm salmon wouldn't escape. When fishermen began catching them, the officials said they couldn't eat wild food. Next they prophesized they would not spawn. Then when juvenile Atlantics were found in the rivers, they said, "Oh well, none of this matters anyway. They won't establish." Why would any but a fool keep on believing these fairy tales? I suspect feral Atlantic salmon cluster outside many a farm, and I suspect DFO agrees because they refuse to send a gillnetter to go look.

Some days I returned to my float with so many Atlantic salmon in my boat I was wading knee-deep through them. I looked like a commercial boat after a night of gillnetting, though my catch was perverted: Atlantic salmon pulled from Pacific waters. One evening I was so exhausted after a day of collecting, measuring, weighing, gutting, preserving stomach contents, and

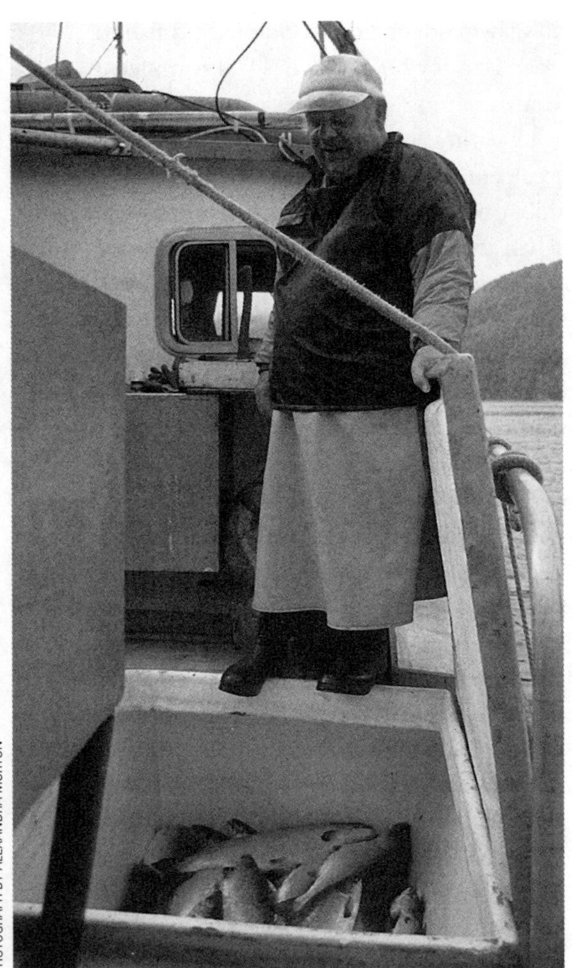

PHOTOGRAPH BY ALEXANDRA MORTON

A perplexed gillnetter ponders his Atlantic catch in BC waters.

taking DNA samples, bacterial swabs, scale samples and pictures that I could not bring myself to deal with the carcasses. As I dragged myself up the ramp, a fleeting thought crossed my mind: "What about the raccoons?" I have a coon problem. A family of these masked bandits climbs into my boat each night and eats everything soft enough to bite, including my daughter's crayons. I knew the next morning there would be farm fish scattered across my float, but I was simply too tired to do anything about it.

The next morning I strode down to the dock, rubber gloves up to my elbows, ready for cleanup detail, but to my surprise there were no fish on the deck. I looked at the situation closely. One Atlantic salmon had been pulled halfway out of the tote and dropped. Its head bore the puncture wounds of a raccoon's bite, but in very short order this raccoon must have decided this was not food. I sat crouched by that fish a long time in disbelief. "It must be the smell of them," I thought, for indeed, I was now very familiar with their smell and they in no way resembled wild fish. They have a cloying odour closely resembling the pellets they eat, which eventually caused me to burn some clothing. Finally I had a good laugh, head back, gasping for air. "My" coons preferred crayons to farm salmon. What would the restaurant crowd think of that?

Partnered Science

Counting, then examining and finally publishing on escaped Atlantic salmon with Dr. John Volpe was an experience that opened a door for me. If DFO wasn't going to do the science, I now knew I could and I would. I would do it for free, because if no one paid me, no one could stop me by withholding funds. This twisted logic was just what the situation needed.

At about this time the internet reached my remote island home, and this changed things. Two-way satellite connections came of age, making my area's lack of a phone system irrelevant. Every time I observed an anomaly I could collect a preliminary smattering of data, find someone who specialized in that field, and start communicating with them. If they became interested I offered to assist them in studying this or took their guidance and did the study myself in silent or stated partnership with them.

This was good. Communicating with bureaucrats was like trying to catch a greased pig—not very productive. Talking with politicians was fun for a while, but I couldn't take their comically furrowed brows seriously anymore. Participating in government processes was designed to absorb useful energy and neutralize it into an inert pudding. I had had enough of that too.

"Partnered science" meant I could go back to what I was good at, watching animals. And this is how I began spiralling down the food chain from orca, the top predator, to fish and finally pathogens, perhaps the most successful life form on earth.

I never quite got over the fish farmers and DFO refusing to reveal information about farmed bacteria in public waters, the same public waters where an entire community was working to bring back wild fish. I was having a bit of a rant about this with Sierra Legal Defence Fund biologist John Werring when he said, "Why don't you test fish yourself?"

"How?" I asked. I had tried this once before, without success. When Bill Proctor got an Atlantic from Scott Cove Creek in 1995, I found a lab to test it for furunculosis. When I phoned the lab for shipping instructions, the man asked, "By the way, what species of salmon is it?"

"Atlantic."

"Oh, well, that's different." He searched for words. "You'll have to send that one to DFO."

"Why? It was caught in a stream."

He was evasive for several minutes and then said, "Look, if I do that fish for you and it has something, I'll never get another job from the DFO or the salmon farmers, and you can appreciate I can't risk that." End of conversation.

But now John told me what to do, and when Bill Proctor caught three more Atlantic salmon in Scott Cove Creek in 1999, I was ready. These fish were all caught days after an escape of Atlantic salmon from the nearby Stolt site at Eden Island. All had three identical sores on them, one on the breast between the pectoral fins and one under each pectoral fin, basically in their armpits.

This time I found a lab outside BC. I didn't go into species specific information, and rather than sending the entire fish I was armed with bacterial swabs. All I had to do was dip one of these oversized Q-tips into the sore, another into the head kidney and then slip them into special tubes where a dollop of jelly would feed the bacteria until they got to the lab.

Everything went smoothly.

When Bill got the second fish I put it on ice in my boat and sped full throttle to Port McNeill to hand it off to the local conservation officer for timely delivery to DFO in Nanaimo.

My samples came back identified as *Serratia liquefaciens.* "Hmmm, that's a new one," I thought. I checked fish disease manuals, talked with various fish pathologists and looked through the hatchery records. Nothing. Then I called my doctor. She knew this bacterium and told me it is found in water that is considered polluted and can infect humans. A researcher in Edinburgh wrote

PHOTOGRAPH BY ALEXANDRA MORTON

An escaped Atlantic salmon caught in 2000.

that *S. liquefaciens* caused a muscle-wasting condition and death of farm salmon, creating lesions like the ones my community was seeing. Furthermore, he went on, an outbreak of this in farmed rainbow trout in Scotland "was associated with leakage from a septic tank."

I recalled one of many anonymous phone calls I received: "I work on a salmon farm, I can't tell you where, but when we flush our toilets, the dye comes up inside the pens." He went on to explain, "When the Ministry of Environment comes around, we stick the longer pipe on, which is supposed to carry the shit past the pens, but the toilet won't actually flush with all that pipe on. I talked to my boss, the MoE, and then my teacher, and she said phone you. I'm afraid we are going to kill someone."

Looking at the lab report, I decided from that moment forward to wear gloves when handling Atlantic salmon. The report on the fish I had given to DFO said the sores were due to *sticks.* We still laugh about that at the hatchery, trying to imagine a salmon getting sticks in both its armpits to create the sores we had all seen.

Atlantics Colonizing the Pacific (2001)

In March 2001 my roving radio ear caught a crab fisherman talking to a friend. When he mentioned he had caught five Atlantic salmon that afternoon in the Wakeman River, I waited impatiently for their conversation to end and then blurted, "Hello, sorry to butt in, but I was just wondering what you did do with those Atlantic salmon?"

Silence. Then a young man answered, "We released them." I wanted to scream, but asked, "You wouldn't be willing to go back there tomorrow and try to catch them again . . . would you?"

Silence.

"Sure."

We met the next day at the Wakeman River, the two crab fishermen coming by seaplane from Echo Bay and I by boat. Steve Vesely, logging camp watchman there for 30 years, met us with his pickup. Steve has been living a long time with grizzly bears and now wears their attitude, powerful and pleasant natured, but not to be messed with. Steve threw his raft single-handed into the pickup despite his more than 70 years on earth, and upriver we jounced. This was fun.

The two fishermen helped me into the raft with my cameras, fish bags and net and we pushed off into winter-clear water. With all the uphill moisture locked in ice there was no silt in the river. The cottonwood were letting loose their intoxicating scent. It was lovely. The guys seemed to share one brain on where each fish had been caught, and so it was amid a constant banter about logs, a certain bush and bends that we alighted on the soft sand bank. Fresh crisp-edged wolf spore trekked across the bar. In seconds fly rods were snaking line out and a fish struck.

"It's the nine-pound buck I hooked here yesterday," and it was.

We leapt back into the raft, stopped and got the twelve-pound doe, then the seven-pound buck and we were done.

The guys flew on to Echo Bay to get back to crab fishing, and I lined up their catch on the floor of my boat. Every animal carries the marks of its life. I study the marks on dolphin fins for clues to what they've been up to, and these fish were no different. One was missing a piece of lower lip as if he'd been hooked before. The doe was rounding out with the development of her eggs. The bucks had wavy parallel lines etched into both sides and the top of their heads. I knew those lines. When dolphins bite each other their evenly spaced teeth scratch lines like these on their victims. But these were much closer together than lines scratched by dolphin teeth; these were the teeth marks of other salmon. These guys had been scrapping. Then I recalled the crab fishermen had not hooked a single wild fish. Was it coincidence or had they been displaced?

These were beautiful fish, coppery with large black spots. When one of the young men was standing hip-deep, hand outstretched, with an Atlantic salmon screaming line off into a wild river, he had turned to me. "Tell me again why this is such a bad thing?" It is every riverman's dream to fight an Atlantic salmon, the king of fish. But the inscriptions on the two male salmon in my boat marked conflict. They had either been fighting each other or another comparably sized fish. John Volpe's work revealed Atlantic salmon and steelhead covet the same territory. Volpe insists he finds Atlantics almost everywhere he looks, and several gillnetters told me the Atlantics they caught hit the same area of the net as steelhead, meaning the two had been swimming together in the marine water. This meant there were a lot of Atlantic salmon getting into rivers, and they were competing with steelhead.

Salmon have a relationship to their river akin to lock and key. Every river runs a little differently; some are faster, glacial, gin-clear, with cobbles or pebbles, sandy, steep, deep, long, short . . . In the ten millennia since the glaciers receded, salmon bodies have been adapting to meet the combination of forces that make each river. Salmon are genetically plastic; each one carries lots of potential variation, and nature makes sure a few salmon always wander off into the wrong river. Each species has a different "straying rate," thus ensuring successful runs will continually send forth emissaries into neighbouring rivers. Only a minute percentage of these explorers will be successful, gradually adapting over millennia to new rivers and building in numbers. For this reason you cannot just throw a pair of salmon into a creek and expect returns on the next cycle. You have to keep trying and keep trying until suddenly you luck out and get a match.

Unlike the Atlantic, where only one very aggressive salmon emerged, the Pacific coast has inspired five and some would argue six or more species of salmon to evolve. Each of these species sends runs home at different times, to different riverine habitat. Chums and pinks spawn the lower intertidal river; coho go into the mountains; and the sockeye need lakes. These different species actually benefit each other; salmon have become a precisely choreographed firework shooting upstream all the ice-free months of the year.

If you drop in another salmon species, one that was not sculpted by time or the crucible of this ecosystem, you literally throw a wrench into the works. Perfectly meshed gears of life can explode. It is well-known that picking up one species and putting it where it did not evolve can wreak havoc, create chaos out of order and destroy biodiversity. As in the natural model, you can get away with numerous introductions of an exotic species that do not take until, shazam, a match. Then all hell breaks loose.

After scrutinizing the faces of the three Atlantics from Wakeman, I opened them up for a deeper look. These fish had been injected with an oil-based vaccine. The males were not all that mature. *Why were they in the river?* But what really caught my attention was the long, puffed, grey-mottled kidney in one. *What do we have here?*

The only people in Echo Bay who know what fish diseases look like are people who have worked on fish farms. So I took my fish to a few ex-fish farmers. They both said the same thing: "Sure looks like BKD." There were references to bacterial kidney disease throughout the publications on salmon farming I was reading. So I got out my handy bacterial swabs, swabbed the fish, packed the swabs to stay cool and express-posted them to a lab on the opposite side of the continent. I always hold my breath when I have bacteria in transit, but they arrived alive.

After confirmation of arrival, I never heard from that lab again even though I still owe it money. I can only guess it was bacterial kidney disease. Once a person starts looking at the little guys, the pathogens, doors slam shut faster than ever.

Algae and Parasitic Palm Trees (2001)

Going boat to boat picking off Atlantic salmon in the summer of 2000 introduced me to most of the local Pacific Management Area 12 salmon fishermen. In August 2001 I took to bobbing at the mouth of Sointula's harbour. There I greeted each fish boat crew as they came in to deliver their catch. The crews came to expect me, and if I could make it easy for them, they were happy to leave their Atlantic salmon on deck and give them to me. As I approached each sleep-starved crew they would wave me over or signal none. When I pulled alongside for a collection, money was sometimes pressed into my hand, and a few "good" fish too, meaning sockeye. "We don't want you eating those pus-bags, Alex," one fisherman admonished. That summer one in five boats had Atlantic salmon onboard, not from a big escape, but picked up scattered throughout the fishing grounds. These fish appeared to be everywhere, making a living on this coast.

In winter, some Sointula gillnetters take their net drums off, thread long poles over their decks and bow, and head off shrimping. My office is a listening station as well as an observatory. I monitor the underwater world with a hydrophone, hearing dolphins, shrimp, cod, boats and very, very rarely now whales. But I also listen to several marine radio channels at a time. This increases my chances of locating whales, as boaters generally talk among each other when they see whales, and so I caught, "*Blackfish Sound, Foxy Lady II,*" one dark January morning.

"Hey Murray, what's up?" I responded

"Ah, we got extraterrestrials here, Alex, attached to the eyeballs of the sole we've been catching. I got some for you if you want them . . . You gotta see this."

How could a woman turn down an offer like that? I headed south through Cramer Passage. There in a bucket were three 18-centimetre arrowtooth sole, each with several palm tree-like things sticking out of their eyes. The fishermen of this area now expect me to figure these things out. Anything with tumours, parasites, open sores is saved for me. I took a picture, scanned it and put it out over the internet. "Does anyone know what this is?"

I got a response from Mississippi, and soon pickled eyeballs with parasitic palm trees *(Phrixocephalus cincinnatus)* were making their way south across North America. The infection rate was off the scale. Generally two percent of any population of arrowtooth sole, also known as turbot, carry this parasite, but here in the Broughton among a cluster of salmon farms over 90 percent had several stuck in each eye. Why? Likely not because the farm salmon were

PHOTOGRAPH BY ALEXANDRA MORTON

Phrixocephalus cincinnatus *on an Arrowtooth sole caught by shrimpers near a salmon farm in the Broughton Archipelago.*

infected with this parasite, but because the massive fecal discharge from each farm had altered the environment so drastically that previous limits to growth had been thrown aside.

Toxic and other algae blooms were staining my home waters red and orange for the same reason. Bill Proctor came over one afternoon. "Christ, the damndest thing happened today," he started. "I was spraying down my hold with seawater when my lips went numb and kinda tingly. Do you think it has anything to do with the tomato soup colour the water has gone around here?"

"Probably."

Salmon farms discharge tonnes of waste each day. Much of this waste is in the form of nitrogen and phosphorous, so algal populations that usually die when they have used up the available nitrogen were exploding on this new, almost limitless diet. Normally this type of plankton response is reserved for coastal areas near large cities or agricultural fields, but Echo Bay had declined to 40 people during the fish farm boom, so it was unlikely we were feeding the plankton. There were, however, 23 salmon farm sites, some containing up to three million fish.

IHN Leapfrogs up the Coast (2001)

In February 2001 my phone started ringing with anonymous callers. "There are three seine boats headed your way out of Vancouver to pump out an entire farm of diseased fish." *Click.* Six hours later, a different voice. "Three seine boats passed Campbell River headed your way." And the next morning. "Nobody in Alert Bay would come out your way to pack diseased fish, so they got three boats from Vancouver. Don't know which farm they are going to." I don't like anonymous calls. I tried to tell the callers that if they had concerns, *they* should speak up, not get me to try and do anything, because so far I was batting zero. I flicked the radio to the farm channels and could hear three boats trying to decide which of two sounds to go to. That told us which farm it was.

It soon became clear this farm had a bad case of IHN, a virus sometimes called "sockeye disease" because sockeye are known carriers. First Nations fishery guardians pulled up as the packers pumped out the diseased salmon and were angered to see what looked like tonnes of herring mixed in with the sick salmon *inside* the pens. They were told that herring swim in when they are small, drawn by the bright night lights used to force rapid growth, and can't get out. As one guardian said to me, "If we did that it would be called poaching. They don't have a licence to harvest herring."

As the seiners left with their highly contagious load, the David Suzuki Foundation and Sierra Legal Defence, working with Vancouver First Nations, did what they could to prevent those boats from bringing the virus into the Fraser River. They were successful in getting an injunction that forbade the seiners to off-load the fish, and the boats wandered for days, trying to find a port that would accept them.

Finally they arrived at French Creek on Vancouver Island and, under the gaze of TV cameras, unloaded the fish to a composting facility.

A few weeks later IHN broke out in salmon farms near Bella Bella, then in Clayoquot Sound. Ultimately 25 percent of the industry became infected with IHN that year (D. Morrison, DVM, Aborig. Fish. Conf., Sept. 2002, author's notes). By this time the network of fishermen, biologists and environmentalists watching salmon farms was so widespread and effective that it was easy to obtain samples of the IHN-infected Atlantics coastwide. I sent the appropriate bits to a lab and learned that the Bella Bella and Broughton strains were identical. Did this mean these fish had come from the same hatchery? Had the farmers stocked their pens with diseased fish again, only this time with a virus rather than bacteria?

I learned the answer when the BC salmon farmers commissioned Campbell River-based Sea to Sky Veterinary Service to do a study on exactly these questions. Sea to Sky reported that in 1999 the virus jumped from wild migrating sockeye to an Omega salmon farm off Kelsey Bay, about 80 kilometres north of Campbell River. Over the winter of 1999–2000 the virus spread to most of the salmon farms in the islands adjacent to Johnstone Strait. When packers carrying live Atlantic smolts to marine netpens passed these farms, they apparently sucked the viral effluent from the farms into their circulating tanks. By the time the packers delivered the young fish to the netpens of the Broughton, Port Hardy and Bella Bella, they were infected with IHN. The report advised packers to turn off the seawater pumps as they passed through this stretch of Johnstone Strait, but of course there was no recommendation on how to protect the Fraser River sockeye that came and went through that area.

Sick fish don't last long in the wild; there are too many predators watching to snap up a wobbling, slow swimmer. But salmon farms keep nature's cleanup crew at bay, allowing fish to die slowly as they shed increasing loads of their infection. This is called "disease amplification," and it is why some scientists have dubbed salmon farms "pathogen culturing facilities."[7] The farms pick up infection from wild fish and then allow it to multiply to levels wild salmon have never encountered before. In the case of IHN, wild sockeye in salt water may carry it, but at levels so low it is hard even for scientists to detect it.

Bill Proctor found the disease record book for the farm where we had first traced the IHN outbreak on a beach. Before returning it he read the chronicle of increasing "morts" (dead fish), beginning with popped eyes in November 2003. The management from there became increasingly dubious. The farm towed a pen of fish past its infected site and into the community picnic grounds on the Burdwood Islands. After the originally infected fish were removed, the virus flared up in the Burdwoods and the farmers apparently decided to gamble and try to rear these highly infectious fish. Mort packers came to and went from that farm from March to November, when they finally sold the last few survivors for human consumption. Early the next spring IHN flared up at a farm nearby. Thousands of fish died faster than three seiners, a packer and a barge could remove them. Decaying bits flooded out of that bay into adjacent Kingcome Inlet in long, snaking, stinking tide-lines. After months of clearing that mess up, the adjacent broodstock site in Cypress Harbour started dying. Two other sites in the Broughton died as well. Mother Nature was taking things into her own hands.

I monitored one of the infected sites frequently that winter. The farm workers did not want me there, but since their boats were roaring in and out, along with fish boats, a log salvage tug and forest company boats, I could not see how my seven metres of fibreglass could possibly add to the bio-security risk they were yelling at me about across the water. As I bore witness, two things struck me as deeply wrong. First, as I peered down through water thick with decomposing Crisco-like fat lumps, and rotting offal, I could see herring flashing and feeding en route to spawn in Kingcome Inlet and Wakeman Sound. In DFO's own studies, 25 percent of herring exposed to IHN became infected and died. These wild fish seemed at grave risk from the farmed epidemic. And second, as tote after tote of putrid-smelling rotting fish were lifted from pens and dumped onto the barge, seine boats were alongside salvaging the living fish out of a soup of corrupted flesh possibly for sale to restaurants.

When BC's minister of Agriculture, Fisheries and Food toured the site, eyeballs and intestines poured under the boat hull and out of the bay from this "quarantined" site. I overheard a government vet remark, "No problem." No problem for WHAT? The fish farmer's deep pockets or the entire 2002 Kingcome Inlet herring stocks and all the species including ours that depend on them? To me it did not look like netpens were working for the Broughton *or* the fish farmers. Six farms near my home went down with IHN in eighteen months, and many others died coastwide, with no end in sight.

Sea Lice and the Crash of the Broughton's Pinks (2001–2002)

On June 5, 2001, the focus of my life altered when my neighbour and friend, sport lodge owner Chris Bennett, tied to my float and strode up my ramp with purpose. "Alex," he blurted out with no preamble, "what do you think these things are all over these salmon fry?" He held up several tiny fish in a jar. It was hard to say. The attachments were small, about the size of sesame seeds and the same yellow-gold colour. The fish were tiny, three-centimetre slips of silver, and there appeared to be about seventeen of the tiny parasites clinging to their sides. The fish did not look healthy. They were emaciated and the telltale redness of infection pooled around the base of their fins.

Juvenile pink salmon infested with sea lice.

"I'll bet you these are baby sea lice," Chris volunteered. "And if they are, this'll be the end of my lodge, everything I have worked for, my dream. This fish is dying. How many more look like this?"

The cords of his neck stood out taut, his face was flushed with rage and fear. He was coming to me because he thought I could figure this out and ultimately stop it. I took the fish. The next day I made a net of crinoline and copper tubing and set out to have a look around.

I doubted I could catch young salmon such as the one Chris had brought. Chris thinks like a fish and can catch anything. As I drew close to the bluffs in Fife Sound I stopped my boat engine and drifted slowly closer. There were thousands of tiny, tadpole-like salmon wriggling parallel to the shore. I raised the net and remained motionless as a blue heron until a school turned toward and under my net. I plunged the net down, collapsing against the boat gunwales as the bow crunched into the rocks, and was surprised to see 25 young salmon in the net. I dumped them into a bucket of water and crouched beside it to have a close look.

Every single fish was bristling with baby sea lice. I went to another spot.

It was the same. I backtracked a few kilometres; more lice. To the north, lice; south, lice. Fortified by my Atlantic salmon and parasitic palm tree studies, I got on the internet, found a listserve for sea lice researchers and typed, "Finding 20–40 sea lice on juvenile salmon 2.7–4.0 cm in length, need help."

Early the next morning came a reply from Norway: "Do you have salmon farms in the area?"

After two days of dialogue the scientist e-mailed: "Dear Alexandra, this is a very controversial subject and I will recommend you not to start a battle against the authorities . . . to be honest I have been waiting for this to come up in your pacific region, here it is accepted this is a problem from fish farms, there is hardly a wild salmon left in the fjords in western Norway (where there are farms)."

I wondered what he was talking about. I wasn't planning to battle the authorities, just study a small louse. With his help and the help of others curious about what, to the best of our knowledge, was the first ever outbreak of these lice on juvenile Pacific salmon, I cobbled together a study and hastily attempted to sample the fish as they headed out to sea.

I travelled in concentric circles from the first sample area, trying to find the edge of the outbreak. Were fish covered with lice the entire length of the BC coast, only here in the Broughton, or only at certain sites in the Broughton? It was like trying to follow water droplets sprayed into a rainstorm. Everything—the water, the fish and the lice—was moving in different directions and at differing rates, and then it was over. A snapshot of a biological anomaly and insidious threat.

The DFO asked me to collect samples for its scientists and then put me under investigation for doing so, threatening that I might have to go to jail. Aha! This was what the scientist from Norway was talking about. The salmon farmers scoffed and ridiculed me. "She can only catch the dying fish." But I counted, sexed and aged thousands of lice on hundreds of fish. Each fish was a planet unto itself, crawling with lice. The lice themselves carried exquisite patterns, like salt crystals, but the fish tore at my heart. They were babies, bleeding—"ravaged" I wrote in my notes.

My partners and I used research done in Norway to start guessing how many lice it would take to kill one of these fish. Pink and chum salmon enter salt water at a size smaller than any other salmon on earth, so none of the work applied directly to them. However, we did know that it takes 1.6 lice per gram of fish to kill a juvenile sea trout. Using this number I calculated how many fish in my sample had been the swimming dead. *Ninety-eight percent* gleamed from the calculator's window.

"Can't be." I refigured the works and the same number glimmered confidently from the little calculator, unaware it was predicting the death of millions. Government and industry started clamoring. "She only caught the sick fish; there are lots of healthy ones down deeper." They sent the huge trawler *Caledonia* and the purse seiner *Odysseus,* caught seven young pink salmon and declared "No problem." I had outfished the Department of Fisheries and Oceans 800 to seven and I knew they were wrong. This was a serious problem.

When that run of pink salmon was due to come home a year and a half later, many were watching for them. Their parents had been the biggest run of pinks into the Broughton Archipelago in recent history. I ran my little *Blackfish Sound* out the archipelago's front door every morning in August, looking for the exuberant leaps of pink salmon coming home. Commercial fishermen with the fish pre-sold watched too. Grizzly mothers nursing cubs watched. Eagles with ravenous youngsters shadowing their every move perched along Knight Inlet watching. Bear-viewing tour operators watched. DFO watched.

First they were forecast, then expected, then they were late and then we realized they were never coming. Ninety-nine percent of the pink salmon from six rivers had vanished. It was one of the biggest recorded salmon crashes since DFO started keeping records in 1957. The eagles dispersed, the grizzlies ate their young, and fishermen wrote editorials. Elsewhere on the coast, pink salmon returns were good to excellent. In the Broughton, three million fish had gone missing in a time of high ocean survival. My forecast of 98 percent annihilation had been dead on.

In the natural system, sea lice die when adult salmon enter a river. They cannot survive fresh water. Therefore, when the young salmon go to sea in the spring, there are no lice to infect them. They cannot get lice until they meet adult lice-carrying salmon out in the ocean, and by then they are large enough to effortlessly carry a few. The vast currents of the oceanic gyres cast lice larvae far and wide. Only the tiniest fraction will find a host. But when a million farm salmon are held stationary in netpens on wild salmon migration routes, the dynamic between parasite and host is reshuffled. The farm salmon pick up lice from adult wild salmon. Instead of dying out over winter, the lice multiply on the farms at a rate that accelerates as the water warms . . . just as the wild babies head to sea. The biology is that simple, though many factors may benefit lice, farmed salmon, or wild fish in alternating cycles. The farms have given lice a super-deluxe inshore habitat for overwintering, which they never

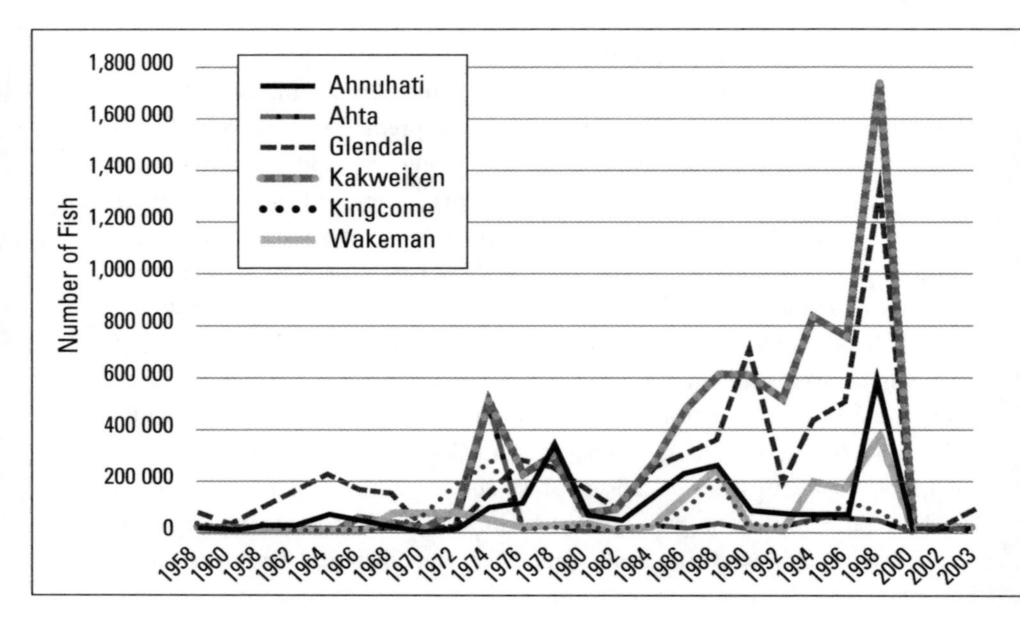

Historic escapement levels in six spawning rivers in the Broughton.

had before.

The DFO sent people to find out why pink salmon had come home fat and numerous at locations from the Arctic to Washington state, but not the Broughton. These DFO scientists calculated that the open ocean hadn't killed the Broughton pinks because other stocks feeding in that same ocean had prospered. They examined the rivers and found this couldn't have been a freshwater calamity because the six rivers that crashed simultaneously had unrelated watersheds. The possibility that mudslides, flooding or heavy poaching had devastated all those rivers was too small to take seriously. That left the nearshore wedge between the rivers and the open ocean, where the babies from the six rivers had travelled. Whatever had befallen these fish, they concluded, had occurred very soon after they left the rivers, in the nearshore marine environment of the Broughton Archipelago.

The Pacific Fisheries Resource Conservation Council (PFRCC), made up of senior DFO biologists headed by John Fraser, the former federal minister of fisheries, met with industry, several non-government groups, DFO, First Nations and me in Campbell River in late fall 2002. There we gave reports, and the evidence indicated the collapse had likely been caused by the sea lice I recorded. The PFRCC leaned toward complete removal of farm salmon from the Broughton. Ready for the fight that always came at these tables in small coastal hotels, my instinct to defend the fish was running high when I realized

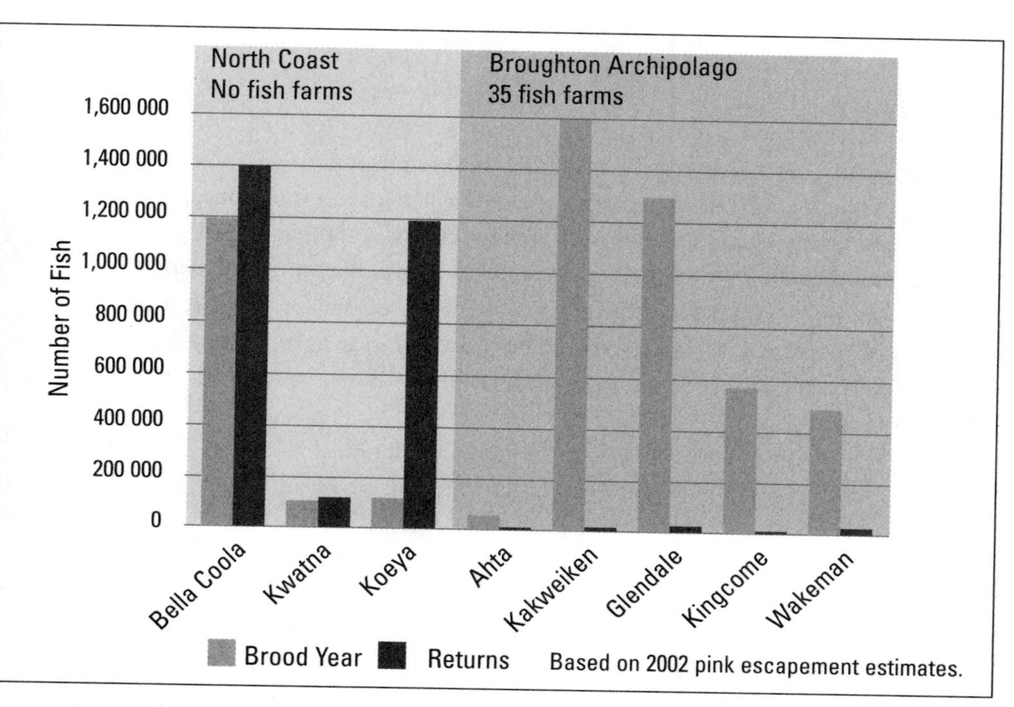

Pink salmon escapement for areas with and without salmon farms.

they were agreeing with me. I was thunderstruck.

I drove the two hours north to Port McNeill with windows down, hair flying and music loud. My heart went singing through those mountains. The DFO had actually looked at the biology of the situation instead of the politics. I was incredulous. A complete removal of the farms in the Broughton! The survivors of this careless extermination would swim to sea unharmed. Oh boy, that felt good. Could I return to studying whales now?

Pink salmon are a gift to every ecosystem lucky enough to have them. Unlike most salmon, pinks do not spend a year feeding in fresh water. In the typically exquisite balance of nature, the carcasses of dead parent salmon feed the insects required by their hatchlings. But in the case of the pink salmon, this abundant fish collects the products of photosynthesis from vast tracts of the open ocean, carries it up mountainsides by the tonne, and then its babies go straight back to sea without partaking in the riches. Pink salmon feed their world. From the moment the soft, translucent eggs leave their mothers' bodies, this new cycle of life feeds birds, wolves, bears, coho, whales, seals . . . their world. In addition, pink salmon are likely the cleanest protein left available to us humans because they feed low on the food chain, avoiding bioaccumulation of the planetary toxins that amass near the top of the food chain, and they only live two years, not long enough to pollute themselves. They are a gift, a world-class food resource, a primary source of energy for their world, a nutrient transport system, the bloodstream of the raincoast.

The triumph of sense began to erode when a second option was birthed a few weeks later. First and best for the wild fish, Option #1 still called for removal of the farm salmon in spring 2003 to give the remaining two percent of the pinks safe passage to sea. But when a second option slipped onto the scene, I instinctively knew it would be chosen. It was vague, hinting at a blend of fallowing some farms, drugs to keep lice levels lower and aggressive monitoring.

"Lowering" lice levels was not going to work unless it was low enough for the pink salmon. As I tried to communicate, the pinks could only live or die. Dying of fewer lice than their parents was no good.

The provincial government picked up the threads, announcing it had a "Sea Lice Action Plan." The pink salmon migration route would be fallowed.

I e-mailed salmon farmers, offering to do anything I could to make sure the public knew they were making a real effort to save the wild fish. They never answered, and then the province disappeared the entire wild fish migration out of Knight Inlet, just as it had disappeared the Ministry of Environment only months before. The government pretended the major migration corridor began at what was really the halfway point.

The farms started counting all sites on the "migration route" that had been emptied for reasons ranging from too much tide to viral epidemics and proclaimed that for the sake of the pink salmon they would fallow eleven farms. "That's okay," I thought. If they were going to make a concession to the wild salmon for the first time since they dipped a toe in these waters, they should be free to embellish.

But then ground started giving way in a familiar pattern. In return for delaying stocking a farm in a migratory red zone, one operator was going to be allowed to start up a new site in a "red" zone where, local fishermen know, millions of Knight Inlet pink salmon would swim past, but this part of the route was not on the provincial map. It did not exist. The operator proclaimed it had spent $200,000 studying the new site, which was said to be, "outside the primary migration route for pinks—local knowledge tells us this, and federal and provincial regulations also tell us this." (*North Island Gazette*, April 9, 2003).

Wild Salmon Force Politicians to Bite All the Hands that Feed Them

I feel I am climbing a shifting mountain of sand. No matter how many times I put one foot ahead of the next, no progress is possible as there is no solid ground. Industry and government appear concerned, appear to make a wise plan, appear to consult with the local yokels, then seem to do whatever the business plan dictates.

The graceful arc of Tribune Channel is like a wind tunnel, switching direction every six hours in response to the ebb and flood tides. Whether you add lice at one end or the other makes no difference; they will blow through the entire system either way. I know this, the provincial biologists know this, DFO knows this because its employees are the ones who did the science, and Stolt knows this because it helped fund DFO to do this science.

As I sit at meeting after meeting where wild things are traded away, I have wondered if this is the moment to stop being reasonable, to leap on the table and invoke spirits, the Lord, bad weather, SOMETHING! As a woman I cannot be shrill. No one listens to shrill women; we are all born with an "off" switch to that. But what? What can I do to turn this tide of short-sightedness?

A fisherman protests in Simoom Sound in the winter of 2003.

Protesters in Simoom Sound, winter 2003.

I can only ask, "Are you sure, absolutely sure, you will *never* want three million delicious, nutritious pink salmon swimming into Tribune Channel again?"

If only I were more than a short, grey-haired woman, could I stop this madness? I feel this is immoral, a genocide of a kind from which we will reel in horror as soon as it is done. What of the children we rob, forevermore, of a gift forged by our planet?

"But Alex," several have started to argue, "this planet has seen mass extinction before, and look, everything is okay."

The malignant power of that statement shatters my mammalian core. Every part of me that is a mother wants to rise in fury. "Do we then choose a planet barren by mass extinction for all who come after us? Is that what we are after here? And for what? So a few can be greedy and sloppy, then move on to raid and loot the next ecosystem until they too realize that *we*, the fish, the rain, buffalo are all one organism. There is no us vs. them, only one. We are one lump of stardust and thus we are killing ourselves, and for what?"

I wish one desperate act could set this straight. But it can't.

I put my video camera into my backpack and point my speedboat toward the new farm site to bear witness. This is all I can think to do. If the salmon farmers are free to do as they please, I am free to document their every move. I focus in on the silvery blue wisps, vibrant newborn lives threading through translucent weaves of kelp. They never stop pouring past my lens, a major seaborne artery, feeding their world as they go. "Fish," I say to my camera, and then tilting the camera to the farm looming on the near horizon I murmur, "Farm." I move up the coast 100 metres. "Fish . . . farm." Cross the channel. "Fish . . . farm." The fish that 200,000 corporate dollars say are not here are in fact everywhere, millions of tiny beating hearts, unmistakably alive. I want to scoop them into buckets and run them out of the Broughton to ensure their safe passage, but I can't. As they swim they are laying down the memory of how to return. We no longer deserve their faith in these waters. Can any reasonable person escape the conclusion that these tender young fish will swim through cloud after cloud of tiny lice wafting out of the farms, a billion tiny lances poised to bury into their flesh and drain their lives away?

Feedlot farming attempts to break immutable laws of nature by overcrowding animals, lowering their genetic diversity and putting them where they do not belong. Nature has a number of tried and true responses to these aberrations. No ecosystem can survive one component going berserk; we call this cancer on a cellular level. Nature will always find a way to reduce a runaway population, effortlessly annihilate genetically identical organisms with a single sweep of pestilence, and vigilantly maintain barriers between ecosystems. Ecosystems are like the succulent bubbles of a pomegranate; they are pressed up against each other, but essential membranes keep them separate. Intact, they have the structure and order to maintain life. Punctured they quickly rot. Nature uses the little guys to enforce these laws, the pathogens—viruses, bacteria and parasites. That farm salmon produce large amounts of sea lice is no great surprise. Everywhere there are salmon farms, wild salmon are dying of sea lice, from Norway through eastern Canada to Supernatural British Columbia.

Despite its imperfections, the provincial Sea Lice Action plan did work to some degree. Even with only half the route cleared of farms, the number of lice per wild salmon fell tenfold. Then I watched the Knight Inlet fish, infected at the east end of the route, collide with the Ahta and Kakweiken fish at the seaward end of the route. Sea lice are a gift that keeps on giving, so the protected fish became infected by their brethren. However, the Ahta and Kakweiken fish had been spared infection for a few weeks. By the time lice settled on them they had doubled in size, and maybe this bought them enough time. Who knows? Salmon farming is a big experiment.

The operator that had promised to delay stocking its site in the migratory red zone did not wait for the wild salmon, even as promised, and restocked the

Glacier Falls site before the migration was over. Meanwhile the previous cycle of adult pinks started rolling home. Once again they were expected, late, diminished . . . Exactly as my second year of lice research predicted, 87 percent of the pink salmon from rivers nearest the farms never returned.

There is no joy in being right about this. My community is reeling. No one among my community has a fish farm job, precious little fish farm dollars are spent here, and postal service has dropped to one day a week during the reign of the salmon farmers. We are a wild fish-based society, and as go the wild fish, so we go. My community is dying of salmon farming. As Eric Nelson, a local stream restoration technician, pointed out, "Why don't the communities profiting from these farms take them into their municipal borders and stop forcing them off on those of us who cannot survive with them?" First Nation neighbours have stated as plainly, clearly and politely as possible that they do not want farms in their territory. "Zero tolerance" they call it as their traditional territory, which they never gave away, is used for a purpose they feel is destroying their way of life. The rage is molten, like lava beneath the surface, and yet there has not been one act of vandalism, violence or anything but willingness to negotiate. But none are heard.

Angry young men charge out at my boat when I go near the farms. They don't want to harass me but have apparently been ordered to chase me away. We both know they have no right to do this. We are in public waters.

"You don't understand," tries one. "I am going to do what it takes to remove you."

I take hold of my VHF radio microphone. "If you touch my boat I am going to issue a mayday and name your company." This is the only warning the person in the smaller boat can make. "Just tell your boss I wouldn't listen to you, that you tried but I was uncooperative . . . Tell him this is not his imaginary kingdom."

The guys laugh, an uneasy truce.

"Why can't I come near your farms? Your lease can't exclude other traffic."

"Because you run all over the place, you're loose."

Hereditary chief Charlie Williams gives me a letter stating I am free to travel throughout his territory. Marine law also gives me "right to innocent passage." So now I exchange documents with the farmers, we shake hands. I will not stop visiting them. If they need to be somewhere private they will have to go where that is possible, but public waters are not the place.

None of this makes sense.

Fish are caught in a southern sea, transported, processed, packaged as feed, transported some more and thrown into a northern sea so fewer fish can be pulled from the sea a second time and sold. I am no business person, so perhaps there is reason to this, but the equation tips further into the red when you calculate the eradication of precious wild stocks. I observed the distortions promoted by government after government: salmon farms will employ local communities—not true in our case; salmon farms are sustainable—not true; salmon farms have minimum impact on wild stocks—not true.

When I first started writing letters, the dull-witted answers gave me the mistaken impression the bureaucratic authors were unintelligent. I kept waiting for them to say, yes, this is a problem, and here is what we are going to do about it, but that answer never came. Instead they tried endless variations on the smoke-and-mirrors theme that might confuse any but those watching from ground zero.

My brain kept turning to the question, What is really going on here? If this was really about raising fish would they maintain diseased fish in the middle of their farm zone? I asked people more experienced then me and finally it became clearer. This must be about oil, logging and water. A wise man said to me, "Alex, wild salmon are politically too expensive, they force politicians to bite all the hands that feed them."

Because wild salmon require functional habitat from the tops of mountains, down through richly forested watersheds, along the coastal shelf and out to sea, politicians can't bear the consequence of taking a stand to protect them. They would have to say "no" to the loggers who want to take the most valuable trees now standing in the last thriving watersheds, "no" to those who scheme to dam, divert and sell BC's fresh water, "no" to miners wanting to dump tailings into the rivers, and most impossibly, "no" to the oilmen greedily eyeing our coast. To these politicians, farm salmon means a salmon that needs no habitat. It is a good deal for them.

It is difficult not to draw a comparison between the East and West coasts of Canada. On the East Coast DFO oversaw the destruction of one of the largest sources of human-grade protein on earth—the cod stocks of the Grand Banks. As observers called out "The cod are declining," DFO invested in a net so large it took two boats to tow it. While cod stocks plunged toward collapse, salmon farms were slipped onto that coast, tossing a few jobs to an economy spiralling into chaos. The final closure of the once-prolific Grand Banks fishery in 1992 roughly coincided with the startup of huge offshore oil projects.

On the West Coast the pesky wild fish are still here, and they are not the only impediment to industrialization, but they are the key one. Whether the irresponsible development of salmon farms in BC is a premeditated strategy for displacing the public fishery resource to clear the way for industry, or whether the mindset that favours industrialization simply breeds indifference to natural values, the result is the same: destruction of the commons by corporate interests, aided by political head-bobbing.

One could ask why this *wouldn't* be happening here.

Epilogue (2004)

As this book is prepared for press, the situation in the Broughton has reached a crucial juncture. Last spring eleven salmon farms were emptied to allow the tiny pink and chum salmon safe passage to sea. This action resulted in much fewer fish infected with sea lice, only 30 percent. But as with so many government initiatives, there was no long-term plan. The DFO produced a paper to prove I was wrong, that the situation could be managed. Last year's play-acting of concern was cancelled and it is back to business as usual this year.

Today Stolt's new Humphrey Rock site is full of farm fish, as are the other three Tribune sites. We had a dry winter, which allowed water salinity to climb to levels sea lice prefer. The collision of corporate and natural forces produced an explosion of lice, and the millions of young pink salmon entering Tribune from Knight Inlet are not swimming out the other end. They will never go to sea.

Every morning I run a transect of Tribune's coastline to pick up the dying. Emaciated and darkened by stress, the fish lie listless on the surface, unaware of my 90 horsepower engine or my hand reaching for them. Below them, schools of stunned pinks and chums are so slow I can park over top of them and stare down at their ruined bodies. The scale of suffering and loss hits deep. I want to scream. I want to save them, but I can only make their deaths public, count and recount and produce the science to describe the carnage.

Salmon farms do not need to make lice and they do not need to kill wild stocks. Unlike logging, where the loggers must take the trees to log, all of this is unnecessary. We are losing the wild salmon of this coast by an act of sheer sloppiness.

As I watch a fish tilt, glinting silver like a falling star, I fear one of our last great food resources is being stolen. Are we supposed to depend on the corporate foodstuffs, give up our access to free nourishment? I know these fish are sacred and should be passed to our children in all their glorious abundance. Do we have what it takes to fight for them? If the next run going to sea in the spring of 2005 is treated to this same holocaust, I fear the Broughton will die. Then I fear the Skeena will die and all the east Vancouver Island rivers and the rivers feeding into the West Coast sounds. One by one I fear they will be taken from us and the destruction will be complete. And when Pacific salmon has finally suffered the fate of the East Coast cod, our spirits will die too. Much better to make a stand now at all costs and restore the soul and source of life on this coast—the wild salmon.

Where Do We Go From Here?

A Fish Farm Action Plan from the David Suzuki Foundation

Continue to ask for wild over farmed salmon in restaurants and grocery stores. For a current list of retailers committed to selling wild salmon, see www.farmedanddangerous.org/solutions.htm. Please contact your local representatives and those provincial and federal ministers responsible for fisheries, environment and aquaculture. For tips and contact information, please see www.davidsuzuki.org/Take_Action.

Governments should be encouraged to do the following:

1. **Remove all open-net cage salmon farms from the marine environment.** Removing open net cages from BC waters and replacing them with closed-loop containment systems prevents waste from being discharged and resolves most environmental issues. All open net cage farms should be converted; farms in wild salmon migration routes or other sensitive areas should be removed.

2. **Remove responsibility for the promotion of aquaculture from Fisheries and Oceans Canada (DFO); increase monitoring and regulation of salmon farming by government regulators.** The department's support and promotion of aquaculture is in conflict with its responsibility to protect wild salmon stocks. Governments should be encouraged to act immediately on the open net cage salmon farming issue. DFO must place a priority on wild salmon and not compromise habitat or stocks for the interests of commercial

aquaculture. Adequate resources should go towards effectively monitoring the industry and enforcing the laws designed to protect our resources and environment.

3. **Increase consultation with communities and First Nations regarding salmon farming operations.** Local communities should be involved in meaningful consultation to avoid conflict and build support.

4. **Reinstate the moratorium on new fish farm sites with no further expansion of existing sites; update the Salmon Aquaculture Review.** Reinstate the BC moratorium and prohibit any further expansion of ocean-based fish farms until significant progress is made on movement to closed containment technology at existing farm sites. Resolving these issues requires updating British Columbia's Salmon Aquaculture Review.

5. **Apply the precautionary approach to regulation of the salmon farming industry.** Regulators should err on the side of caution to protect environmental values, wild fish, and productive fish habitat. The importance of the values and the resources at stake requires that the precautionary approach be applied.

6. **Require labelling and identification of farmed fish at the consumer level.** Farmed fish should be identified distinctly from wild fish in retail outlets and restaurants so consumers can make informed choices. Farmed fish is often labeled "fresh," or, in the case of salmon as "Atlantic." For many consumers, the relevant distinction is "farmed" versus "wild."

7. **Move away from farming carnivorous species and towards the farming of herbivorous species.** As long as carnivorous (piscivorous) fish species, such as salmon, cod and halibut are being farmed, there will be an overall world protein loss. Often the "feed-fish" being caught to feed first world farmed fish come from developing nations where the fish represent an important food source. Farming herbivorous fish species, such as tilapia and carp, would add to, rather than lessen, the global food supply.

These recommendations were originally published in *Clear Choices, Clean Waters: The Leggatt Inquiry into Salmon Farming in British Columbia* by the Honourable Stuart M. Leggatt, 2001. They have been updated for this book by the David Suzuki Foundation's Marine Conservation Program.

Contributors

Stephen Hume, a columnist for the *Vancouver Sun*, has won more than a dozen awards for his poetry, essays and journalism and is the author of eight books, including *Bush Telegraph* and *Off The Map*. In 2001, Stephen became the first Canadian to win a Dolly Connelly prize for environmental writing.

Betty C. Keller is an author, playwright, and editor who lives on the Sunshine Coast of BC. A founder of the Festival of the Written Arts in Sechelt, BC, she is also the author of numerous books, including *Forests, Power and Policy* (with Eileen Williston) and the novel *Better the Devil You Know*.

Rosella M. Leslie is the author of *The Sunshine Coast: A Place to Be*, and co-author with Betty Keller of *Bright Seas, Pioneer Spirits* and *Sea Silver: Inside British Columbia's Salmon Farming Industry*. She was born in Edmonton, Alberta and now lives in Sechelt, BC.

Otto Langer, a former federal Department of Fisheries and Oceans biologist, is now the Director of the Marine Conservation Program for the David Suzuki Foundation and one of DFO's most outspoken critics. Langer is considered one of Canada's leading authorities on the issue of open net cage salmon farming.

Don Staniford is a Director of the Salmon Farm Protest Group in Scotland. He has spoken in opposition of salmon farming in Brussels, Chile, Australia and New Zealand. In 2002, he received a British Environment and Media Award in recognition of his work exposing the illegal use of chemicals on Scottish salmon farms.

Alexandra Morton moved to the BC coast in 1979 to study the orcas that frequented the Johnstone Straits and Broughton Archipelago. In the 1990s, she began to study salmon farming. Morton has authored seven academic papers and published five books, including *Heart of the Raincoast* with Bill Proctor. She lives in Echo Bay, BC.

Notes

Sea-Silver: A Brief History of British Columbia's Salmon Farming Industry

1. At different times this ministry was known as Agriculture; Agriculture and Food; Agriculture and Fisheries; Agriculture, Fisheries and Food; Agriculture, Food and Fisheries; and it is referred to by various names throughout this essay. In the early years, fisheries was under the jurisdiction of the Ministry of Recreation and Conservation.
2. Department of Fisheries and Oceans, *Federal Aquaculture Development Strategy* (Ottawa: DFO, 1995), pp. 7, 13.
3. Robert Williamson, "BC Fish Farmers Look Elsewhere," *Globe and Mail,* March 27, 1995, pp. B1 and B8.
4. Telephone interview with Julian Thornton, November 30, 1995.
5. Telephone interview with David Groves, March 23, 1995.
6. *Sunshine Coast News,* February 26, 1985.
7. Interview with Reid Arnold on May 10, 1995.
8. Marianne Holmer and Erik Kristensen, "Impact of Marine Fish Cage Farming on Metabolism and Sulphate Reduction of Underlying Sediments," *Marine Ecology Progress Series* 80 (1992): 191–201.
9. J.G. Stockner, D.D. Cliff and K.R.S. Shortread, "The Phytoplankton Ecology of the Strait of Georgia, British Columbia," *Journal of Fisheries Resource Board, Canada* 36 (1979).
10. Interview with Dr. F.J.R. "Max" Taylor, June 8, 1995.
11. *Sunshine Coast News,* July 16, 1985.
12. Mrs. D.E. McTaggart, Letter to the Editor, *Sunshine Coast News,* September 23, 1985.
13. *Sunshine Coast News,* May 12, 1986.
14. Telephone interview with Ron Ginetz, May 10, 1995.
15. *Sunshine Press,* February 27, 1986.
16. *Aquaculture on the Sunshine Coast: A Review by the Ministry of Lands, Parks and Housing* (Victoria: MLPH, n.d.), p. 1.
17. Telephone interview with Edward Black, November 1995.
18. Interview with Dr. Max Taylor, June 8, 1995.
19. David Gillespie, *An Inquiry into Finfish Aquaculture in British Columbia* (Victoria: Government of BC, 1986), p. 37.
20. Telephone interview with Bjorn Skei, October 22, 1995.
21. *Fish Farm News* 2, no. 3 (March 1989): 15.
22. Rob Russell to Paul Sprout, August 11, 1988.
23. *Sunshine Press,* April 25, 1989, p. 18.

24. Ibid.
25. *Vancouver Sun,* July 30, 1988.
26. Ibid.
27. *Fish Farm News* 2, no. 5 (May 15, 1989).
28. *Fish Farm News* 2, no. 8 (August 7, 1989).
29. Telephone interview with Ward Griffioen, October 12, 1995.
30. *Fish Farm News* 3, no. 9 (September 1, 1990): 7.
31. Telephone interview with Sandy Brook, October 11, 1995.
32. Interview with Dr. Max Taylor, June 8, 1995.
33. Telephone interview with Greg Rebar, November 19, 1995.
34. *Fish Farm News* 5, no. 7 (September 1992): 10.

Interviewees

BC Ministry of Agriculture, Fisheries and Food

Clare Backman, finfish biologist
Edward Black, specialist, Aquatic Resources
Dr. Joanne Constantine, fish health veterinarian for Courtenay
Mike Coon, manager, Marine Resource Planning
Rick Deegan, information officer
Bill Harrower, marine finfish extension biologist

BC Ministry of Environment, Lands and Parks

Doug Berry, land officer
Don Peterson, manager, Fish Culture Section

Department of Fisheries and Oceans

Barbara Aweryn, senior chemist, DFO Chemistry Laboratory
Dr. Craig Clark, head of Fish Culture Research, Pacific Biological Station, Nanaimo
Dr. Robert H. Devlin, head of Molecular Biology Program, West Vancouver Laboratory
Dr. Edward M. Donaldson, West Vancouver Laboratory
Dr. T.P.T. Evelyn, head of Fish Health and Parasitology Division, Pacific Biological Station, Nanaimo
Kelly Francis, communications officer, Pacific Biological Station, Nanaimo
Ron Ginetz, chief of aquaculture
Dr. David Higgs, head of Nutrition Program, West Vancouver Laboratory
Dorothee Kieser, biologist, Fish Division
Grant McBain, field officer, Salmon Enhancement Program
Peter Olesiuc, biologist, Marine Mammal Section
Ted Perry, acting chief for Program Coordination Assessment Division, Salmon Enhancement Program
Greg Steer, evaluation coordinator, Salmon Enhancement Program
John Yarish, technical services supervisor, Inspection Division, Quality Managment Program

Researchers

Dr. Lauren Donaldson, University of Washington, Aquaculture Resources
Dr. Dianne Elliott, US National Biological Service
Dr. Lee Harrell, National Marine Fisheries, Manchester Experimental Fish Farm
Dr. F.J.R. "Max" Taylor, Department of Oceanography, University of British Columbia
Craig Steveson, Centre for Coastal Health, University of British Columbia
Ron Fearn, Sunshine Coast teacher
Steve Marsh, formerly with Capilano College Aquaculture Resource Centre
Alexandra Morton, Raincoast Research

Industry

Kjell Aasen, Liard Aquaculture Ltd.
Reid Arnold, formerly with Bayfresh Farms Ltd., Salmon Pacific Salmon Inc., Rondo Seafarms Ltd. and Finstar Seafarms ltd.
Bernie Bennett, Target Marine Products Ltd.
Dale Blackburn, Stolt Sea Farm Inc.
Sam Bowman, Moore-Clark Company (Canada) Ltd.
Sandy Brook, formerly with Western Harvest Seafoods
Roger Engeset, Wood Bay Salmon Farms
Joyce Francis, Target Hatcheries Ltd.
Dr. Don French, formerly with Moore-Clark Company (Canada) Ltd.
Sheryl French, Blue Tornado Enterprises Ltd.
Dan Gillis, formerly with Nimpkish Indian Band Seafarm
Dora Glover, Sunshine Coast Aquaculture Association
Doug Goodbrand, formerly with Indian Arm Salmon Farm Ltd.
Ward Griffioen, West Coast Fishculture (Lois Lake) Ltd.
Dr. David Groves, Seaspring Salmon Farm Ltd.
Syd Heal, Sunshine Coast Aquaculture Association
Dr. John Heath, Yellow Island Aquaculture Ltd.
Brad Hicks, International Aqua Foods Ltd.
John Hutchison, formerly with Tonto Seafarms Ltd.
Dr. Grace Karenan, Cooperative Assessment of Salmonid Health, BC Salmon Farmers Association
William Luers, formerly with Crown Zellerbach Ltd.
Linda May, formerly with Royal Pacific Seafoods Ltd.
Colin McMillan, J.S. McMillan Fisheries Ltd.
Ted Needham, BC Packers Ltd.
F. "Pete" Petereit, formerly with Crown Zellerbach Ltd.
Mrs. Percy Priest, formerly with Alberni Marine Farms
Gary Pullen, Pacific National Group
Greg Rebar, formerly with IBEC
John Ronnekleiv, Rondo Seafarms Ltd. and Anchor Seafarms Ltd.

W.A. "Bill" Sharkey, retired Crown Zellerbach Ltd. executive
Barb Sharpe, Quartz Bay Sea Farms Ltd.
Bjorn Skei, Sechelt Salmon Farms Ltd.
John Slind, formerly with Suncoast Salmon Ltd.
Mavis Smeal, Saltstream Engineering Ltd.
Dr. Julian Thornton, Microtech International Ltd.
Jon Van Arsdell, formerly with Moccasin Valley Marifarms
Bill Vandevert, West Coast Fishculture (Lois Lake) Ltd.
Jack Waterfield, Lions Gate Fisheries

Silent Spring of the Sea

1. R. Carson, *The Sea around Us* (Oxford University Press: New York, 1951); R. Carson, *The Edge of the Sea* (Oxford University Press: New York, 1955); R. Carson, *Silent Spring* (Penguin Books: London, 1962).
2. J. Hamilton-Paterson, *Seven-Tenths: The Sea and Its Thresholds* (Vintage: London, 1993).
3. D. Staniford, "The Five Fundamental Flaws of Sea Cage Fish Farming" (paper presented at the European Parliament's public hearing in Brussels on Aquaculture in the European Union, October 1, 2002), http://www.watershed-watch.org/ww/publications/sf/ Staniford_Flaws_SeaCage.PDF.
4. World Health Organization, "Food Safety Issues Associated with Products from Aquaculture," WHO Technical Report Series 883 (Geneva: WHO, 1999), http://www.who.int/foodsafety/publications/ fs_management/aquaculture/en/.
5. R.A. Hites et al., "Global Assessment of Organic Contaminants in Farmed Salmon," *Science* 303 (2004): 226–229, http://www.albany.edu/ ihe/salmonstudy/index.html.
6. M.J. Mac et al., "PCBs and DDE in Commercial Fish Feeds," *Progressive Fish Culturist* 41 (1979): 210–211; M. Jacobs et al., "Organochlorine Pesticide and PCB Residues in Pharmaceutical, Industrial and Food Grade Fish Oils," *International Journal of Environment and Pollution* 8, no. 1–2 (1997): 74–93; K. Oetjen and H. Karl, "Levels of Toxaphene Indicator Compounds in Fish Meal, Fish Oil and Fish Feed," *Chemosphere* 37, no. 1 (1998): 1–11; UK Ministry of Agriculture Fisheries and Food, "Dioxins in Fish and Fishery Products" (London: MAFF, 1999), http://www.foodstandards.gov.uk/maff/archive/food/infsheet/1999/no184/184diox .htm; M. Jacobs et al., "Investigations of CDDs, PCDFs and Selected Coplanar PCBs in Scottish Farmed Atlantic Salmon," *Organohalogen Compounds* 47 (2000): 338–341. A.K. Lundebye et al., "Documenting Seafood Safety: Contaminant Concentrations in Norwegian Fish Feeds and Mariculture Products," ICES CM P:03 (Copenhagen: International Council for the Exploration of the Sea, 2000): 1–2; M.D.L. Easton et al., "Preliminary Examination of Contaminant Loadings in

Farmed Salmon, Wild Salmon and Commercial Salmon Feed," *Chemosphere* 46 (2002): 1053–1074, http://www.elsevier.com/inca/publications/store/3/6/2/; M. Jacobs, "Persistent Organic Pollutants in Fatty Fish—Issues and Concerns" (paper presented at Aquavision 2002, June 12, 2002), http://www.aquavision.nu; M. Jacobs et al., "Investigations of Polychlorinated Dibenzo-p-dioxins, Dibenzo-p-furans and Selected Coplanar Biphenyls in Scottish Farmed Salmon," *Chemosphere* 47, no. 2 (2002): 183–191; M. Jacobs et al., "Investigation of Selected Persistent Organic Pollutants in Farmed Atlantic Salmon, Salmon Aquaculture Feed, and Fish Oil Components of the Feed," *Environmental Science and Technology* 36, no. 13 (2002): 2797–2805.

7. J.E. DuBois, *The Devil's Chemists* (Boston: Beacon, 1952); S. Tvedten, *The Best Control: Intelligent Pest Management* (Published by the author, 2001), http://www.safe2use.com/ca-ipm/01-04-27.htand http://www.thebestcontrol.com/toc/toc1-10.htm.
8. C. Sommerville, "Latest Weapons in the War on Sea Lice," *Fish Farmer* 18, no. 2 (1995): 53–55; J.A. Mackie, "Fighting the Number One Enemy," *Scottish Fish Farmer* 95 (1996): 16–17; "Chemical Warfare," *Scottish Fish Farmer* (May 1997).
9. J.G. McHenery et al., "Threshold Toxicity and Repeated Exposure Studies of Dichlorvos on the Larvae of the Common Lobster," *Aquatic Toxicology* 34 (1996): 237–251; J.G. McHenery et al., "Experimental and Field Studies of Effects of Dichlorvos Exposure on Acetlycholinesterase Activity in the Gills of the Mussel, *Mytilus edulis L.*," *Aquatic Toxicology* 38 (1997) 125–143; D.J. Murrison et al., "Epiphyte Invertebrate Assemblages and Dichlorvos Usage at Salmon Farms," *Aquaculture* 159 (1997): 53–66; S. McKeown and S.J. Hay, "A Preliminary Investigation of the Toxicity and Sub-lethal Effects of the Sea Lice Control Chemicals Dichlorvos and Ivermectin on Survival and Feeding of the Marine Copepod *Acartia tonsa*," Marine Laboratory Report No. 9/98 (Aberdeen: Scottish Marine Laboratory, 1998); O. Tully and Y. McFadden, "Variation in Sensitivity of Sea Lice to Dichlorvos on Irish Salmon Farms in 1991–92," *Aquaculture Research* 31, no. 11 (2000): 849–854.
10. K.J. Willis and N. Ling, "The Toxicity of Emamectin Benzoate, an Aquaculture Pesticide, to Planktonic Marine Copepods," *Aquaculture* 221 (2003): 289–297; K.J. Willis and N. Ling, "Toxicity of the Aquaculture Pesticide Cypermethrin to Planktonic Marine Copepods," *Aquaculture Research* 35 (2004): 263–270.
11. Institute of Aquaculture Stirling, "An ERA of the Use of Teflubenzuron to Control Ectoparasite Infestations on European Salmon Farms" (confidential report for Nutreco, 1998); Institute of Marine Research, "Tolerance of Juvenile Lobsters to a Feed Additive for Oral Treatment of Salmon Lice on Atlantic Salmon" (confidential report to Nutreco, 1995); Nutreco, "Long Term Environmental Monitoring of Teflubenzuron Used for the Treatment of Sea Lice in the Marine Environment," Addendum I and II to the Interim Report ARC-TFBZ-UK-5-98

(confidential report for Nutreco, 1998); Nutreco, "Environmental Risk Assessment of a Nutreco Insecticide" (confidential report for Nutreco, 1998); G. Ritchie, "Long Term Environmental Monitoring of Teflubenzuron Used for the Treatment of Sea Lice in the Marine Environment" (confidential report to Nutreco, 1999).

12. "Study reference GP95033" (report from Grampian Pharmaceuticals Ltd. Research Division, 1995), 76; J.G. McHenery et al., "Toxicity of FD2 to Larvae of the Common Lobster," Fisheries Research Services Report 9/91 (private and confidential report, 1991); G. Ritchie, "Efficacy and Action of CME-134 Used as an Oral Treatment for the Control of Sea Lice," *Bulletin of the Aquaculture Association of Canada* 4 (1996): 26; T.C. Telfer, "Marine Environmental Effects Monitoring and Dispersion Study of the Anti-Sea Lice Drug SCH5884 in Loch Duich under Commercial Use Conditions" (Schering Plough Animal Health Study No. 1090N-60-V97-357, 1998).
13. Ciba Geigy, "Nuvan Fish 500 EC. Product Licence Submission Supporting Data," Volume 5: Environmental Impact (private and confidential report, 1987); A.L.S. Munro, "Use of Nuvan Fish 500 EC in Salmon Farms and Consideration of its Environmental Impact" (private and confidential report for Ciba Geigy, 1987); D.P. Dobson, "Nuvan Fish 500 EC: Loch Dispersion Field Trial" (private and confidential report for Ciba Geigy, 1988); J. McHenery, "Acute Toxicity of Nuvan Fish 500 EC to Larvae of *Homarus gamarus L* and *Clupea harengus L*" (private and confidential report for Ciba Geigy, 1988); A.L.S. Munro, "DAFS Marine Laboratory Response to Ciba-Geigy Product Licence Application for Nuvan 500 EC" (private and confidential report for Ciba Geigy, 1988).
14. UK Parliamentary Debates, Commons, February 21, 1989 (Available online at http://www.parliament.the-stationery-office.co.uk).
15. I.M. Davies et al., "Effects of TBT in Western Coastal Waters" (private and confidential final report to DETR, contract PECD CW0691: Fisheries Research Services Report No. 5/98, not available from the Scottish Executive, 1998); J. Duffus, "An Environmental Impact Assessment with Regard to the Possible Use of Azamethiphos to Control Sea-Lice in Salmon" (confidential report prepared for Ciba-Geigy Agriculture, 1992); R.P. Hunter and N. Fraser, "Field Monitoring of the Effects of Cypermethrin as GPRD01" (confidential report of Grampian Pharmaceuticals Ltd., Research Division, Ref GP95033, 1995); P.H. Rose et al., "Canthaxanthin Potential Tumorigenic and Toxic Effects in Prolonged Dietary Administration to Mice" (unpublished report HLR 135/861058, submitted to WHO by Hoffmann-La Roche, Basle, 1987); J.G. McHenery, "Review of Environmental Data Relating to the Use and Disposal of Cypermethrin Formulated as GPRDO1 (Excis) as a Bath Treatment for Salmon" (Inveresk Research International Project 384259, 1996); J.G. McHenery, "Assessment of the Potential Environmental Impacts of the Use of Ivermectin as an In Feed Medication for Salmon" (private and confidential Inveresk Research International

report for the Scottish Salmon Growers Association, 1996); J.G. McHenery and C.M. Mackie, "Revised Expert Report on the Potential Environmental Impacts of Emamectin Benzoate, Formulated as Slice, for Salmonids," Cordah Report No. SCH001R5 (confidential report to Schering-Plough Animal Health, 1999); T. Needham, "Elimination of the Salmon Louse Using DDVP—Nogos 50" (Unilever Research Summary Report. Private and Confidential, 1978); T. Nickell, "The Effects of Emamectin Benzoate on Infaunal Polychaetes," Final Report DML Project 20898 for Schering Plough Animal Health (Dunstaffnage Marine Laboratory Internal Report 226, 2002); W. Köpcke et al., "Canthaxanthin Deposition in the Retina: A Biostatistical Evaluation of 411 Cases Taking this Carotenoid for Medical or Cosmetical Purposes" (unpublished manuscript submitted by Hoffman-La Roche, Basle, 1992, http://europa.eu.int/comm/food/fs/sc/oldcomm7/out10_en.html); R.R. Stephenson, "The Acute Toxicity of Cypermethrin (WL 43467) to the Freshwater Shrimp and Larvae of the Mayfly in Continuous Flow Tests" (confidential Shell Research Report No. TLGR-80.079, Sittingbourne, 1980).

16. "SEPA to Speed Sea Lice Treatment Process" (release from Scottish Environment Protection Agency, February 19, 2002), http://www.sepa.org.uk/news/releases/2002/pr026.html.

17 OSPAR, "PARCOM Recommendation 94/6 on Best Environmental Practice for the Reduction of Inputs of Potentially Toxic Chemicals from Aquaculture Use" (report from Oslo and Paris Conventions for the Prevention of Marine Pollution, 16th joint meeting, 1994); Scottish Executive, "PARCOM Recommendation 94/6 on Best Environmental Practice (BEP) for the Reduction of Inputs of Potentially Toxic Chemicals from Aquaculture Use: United Kingdom Code of Best Environmental Practice," 2000.

18. "Aquaculture Chemicals: BC Government Ignores Their Own Law—The Unpermitted Use of Powerful Pesticides to Treat Sea Lice Violates Provincial Pesticide Control Act" (release from Public Service Employees for Environmental Ethics, March 6, 2003), http://www.pse.ca/ march6release.html.

19. "Environmental Objectives for Norwegian Aquaculture: New Environmental Objectives for 1998–2000" (Trondheim, Norway: Directorate for Nature Management, 1999), http://193.217.72.207/ filer/pdf/Aquaculture.pdf.

20. "Contaminated Chilean Salmon Impounded in Europe," *Salmon Farm Monitor,* August 2003, http://www.salmonfarmmonitor.org/intlnewsaugust2003.shtml#item1; "Malachite Green Contamination in Chilean Salmon," *Salmon Farm Monitor,* September 2003, http://www.salmonfarmmonitor.org/intlnewsseptember2003.shtml#item1.

21. "Fishy Business Exposed in BC Aquaculture—Sierra Legal Defence Fund Demands Investigation of Suspicious Dealings Regarding Fines against Aquaculture Companies," Sierra Legal Defence Fund, February 5, 2004,

http://www.sierralegal.org/m_archive/pr04_02_05.html.

22. G. Rhodes et al., "Distribution of Oxytetracycline Resistance Plasmids between Aeromonads in Hospital and Aquaculture Environments: Implications of TN1 721 in Dissemination of the Tetracycline Resistance Deteriminant Tet A.," *Applied Environmental Microbiology* 66, no. 9 (2000): 3883–3890; D.J. Alderman and T.S. Hastings, "Antibiotic Use in Aquaculture: Development of Resistance—Potential for Consumer Health Risks," *International Journal of Food Science and Technology* 33, no. 2 (1998): 139–155; J.R. MacMillan, "Aquaculture and Antibiotic Resistance: A Negligible Public Health Risk?" *World Aquaculture* June 2001; C. Benbrook, "Antibiotic Drug Use in U.S. Aquaculture" (Minneapolis: Institute of Agriculture and Trade Policy, 2002), http://www.iatp.org/fish/library/uploadedFiles/Antibiotic_Drug_Use_in_US_Aquaculture.doc; O.B. Samuelson et al., "Residues of Oxolinic Acid in Wild Fauna Following Medication in Fish Farms," *Diseases of Aquatic Organisms* 12 (1992): 111–119; A. Ervik et al., "Impact of Administering Antibacterial Agents on Wild Fish and Blue Mussels in the Vicinity of Fish Farms," *Diseases of Aquatic Organisms* 18 (1994): 45–51; J. Kerry et al., "Spatial Distribution of Oxytetracycline and Elevated Frequencies of Oxytetracycline Resistance in Sediments beneath Marine Salmon Farms Following Oxytetracycline Therapy," *Aquaculture* 145, no. 1–4 (1996): 31–39; D.G. Capone et al., "Bacterial Residues in Marine Sediments and Invertebrates Following Chemotherapy in Aquaculture," *Aquaculture* 145 no. 1–4 (1996): 55–75; P.G. Provost et al., "Antibiotics in Fish Farm Sediments," in *Environment Pollution: Assessment and Treatment,* ed. P. Read and J. Kinross (Edinburgh: Napier University Press, 1997).
23. "Report of the Ad Hoc Study Group on 'Environmental Impact of Mariculture'" (Copenhagen: International Council for the Exploration of the Sea, March 1988); "Environmental Objectives for Norwegian Aquaculture: New Environmental Objectives for 1998–2000; T. Horsberg, "Food Safety Aspects of Aquaculture Products in Norway" (release from Atlantic Institute for Market Studies, Halifax, NS, 2000), http://www.aims.ca/Aqua/horsberg.htm; A. Lillehaug et al., "Epidemiology of Bacterial Diseases in Norwegian Aquaculture: A Description Based on Antibiotic Prescription Data for the Ten-year Period 1991 to 2000," *Diseases of Aquatic Organisms* 53, no. 2 (2003): 115–125, http://www.ncbi.nlm.nih.gov/entrez/query.fcgi?cmd=Retrieve&db=PubMed&list_uids=12650244&dopt=Abstract.
24. "Environmental Objectives for Norwegian Aquaculture: New Environmental Objectives for 1998–2000." Norwegian Directorate for Nature Management, 1999: http://www.dirnat.no/wbch3.exe?ce=3118.
25. F. Cabello, "Antibioticos y Acuicultura: Un Analysis de sus Potenciales Impactos para el Medio Ambiente, la Salud Humana y Animal en Chile," (Santiago: Terram, 2003), http://www.terram.cl; D. Hide, "Salmon: A Threat from Chile?" *The Farmers Club Journal,* Autumn 1996; "Chile Caught Using 75 Times More Antibiotics Than

Norway," *Salmon Farm Monitor,* July 2003, http://www.salmonfarmmonitor.org/intlnewsjuly2003.shtml#item10; C.D. Mirand and R. Zemelman, "Antimicrobial Multiresistance in Bacteria Isolated from Freshwater Chilean Salmon Farms," *Science of the Total Environment* 293, no. 1–3 (2003): 207–218, http://www.ncbi.nlm.nih.gov/entrez/query.fcgi?cmd=Retrieve&db=PubMed&list_uids=12109474&dopt=Abstract; "Japan to Ban Chilean Farmed Salmon?" *Salmon Farm Monitor,* October 2003, http://www.salmonfarmmonitor.org/intlnewsoctober2003.shtml#item2.

26. G.M. Lalumera et al., "Preliminary Investigation on the Environmental Occurrence and Effects of Antibiotics Used in Aquaculture in Italy," *Chemosphere* 54, no. 5 (2004): 661–668, http://www.ncbi.nlm.nih.gov/entrez/query.fcgi?cmd=Retrieve&db=PubMed&list_uids=14599512&dopt=Abstract; A.S. Schmidt et al., "Occurrence of Antimicrobial Resistance in Fish-pathogenic and Environmental Bacteria Associated with Four Danish Rainbow Trout Farms," *Applied Environmental Microbiology* 66, no. 11 (2000): 4908–4915, http://www.ncbi.nlm.nih.gov/entrez/query.fcgi?cmd=Retrieve&db=PubMed&list_uids=11055942&dopt=Abstract; H. Lutzhoft, S. Halling and S. Jorgensen, "Algal Toxicity of Antibacterial Agents Applied in Danish Fish Farming," *Archives of Environmental Contamination and Toxicology* 36, no. 1 (1999): 1.
27. "Final Report of the MPMMG Subgroup on Marine Fish Farming," Scottish Fisheries Working Paper No. 3 (Aberdeen: Scottish Office, 1992).
28. J.W. Treasurer et al., "Physical Constraints of Bath Treatments of Atlantic Salmon with a Sea Lice Burden," *Contributions to Zoology* 69 (2000): 129–136.
29. A.N. Grant, "Medicines for Sea Lice," *Pest Management Science* 58 (2002): 521–527.
30. A. Berge, "Salmon Farming Rivals 2003," Intrafish, February 2004, http://www.intrafish.com/intrafish-analysis/.
31. Norwegian Directorate of Fisheries, "Key Figures from the Norwegian Aquaculture Industry, 2000" (Norway: Department of Agriculture, 2001), http://www.fiskeridir.no/english/pages/statistics/ keyfigures_aqua_00.pdf; Statistics Norway, "Amount of Pharmaceuticals Sold for Use in Fish Farming (some substances): Agents Used against Endoparasites/Ektoparasites (1989-2001)" (report from Statistics Norway, 2003), http://www.ssb.no/english/subjects/10/05/nos_fiskeoppdrett_en/nos_d259_en/tab/4.2.html.
32. D.J. Alderman, "Chemicals Used in Mariculture," Cooperative Research Report No. 202 (Copenhagen: International Council for the Exploration of the Sea, 1994); D.J. Alderman, "Chemicals in Aquaculture," in *Sustainable Aquaculture* (Rotterdam: Balkema, 1999); D.J. Alderman and C. Michel, *Chemotherapy in Aquaculture: From Theory to Reality* (Paris: Office International des Epizooties, 1992); J.F. Burka et al., "Drugs in Salmonid Aquaculture: A Review," *Journal of Vet. Pharmocol. Therap.* 20 (1997): 333–349; Conservation Council of New Brunswick, "Pesticides in Salmon Aquaculture in Southwest New Brunswick" (a background paper prepared for

World Wildlife Fund Canada, Toronto, 1998); M.J. Costello et al., "The Control of Chemicals Used in Aquaculture in Europe," *Journal of Applied Ichthyology* 17, no. 4 (2001): 173–180; GESAMP "Towards the Safe and Effective Use of Chemicals in Coastal Aquaculture," Report No. 65 (Rome: Food and Agriculture Organization, 1997), http://gesamp.imo.org/no65/index.htm; Grant, "Medicines for Sea Lice"; T. Kasa, "Consumption of Chemicals in Norwegian Aquaculture" (SFT Report No. 91, Norway, 1991); B.T. Lunestad, "Therapeutic Agents in Norwegian Aquaculture, Consumption and Residue Control" (report for the National Institute of Nutrition and Seafood Research, Norway, 2001); F.P. Meyer and R.A. Schnick, "A Review of Chemicals Used for the Control of Fish Diseases," *Review of Aquatic Sciences* 1 (1989): 693–710; I. Milewski, "Impacts of Salmon Aquaculture on the Coastal Environment: A Review" (paper presented at a conference in New Brunswick, Canada, 2001), http://www.iatp.org/fish/library/uploadedFiles/Impacts_of_Salmon_Aquaculture_on_the_Coastal_E.pdf; "Aquaculture Chemicals: BC Government Ignores Their Own Law—The Unpermitted Use of Powerful Pesticides to Treat Sea Lice Violates Provincial Pesticide Control Act" (release from Public Service Employees for Environmental Ethics, March 13, 2003), http://www.pse.ca; "Pesticide Use in Aquaculture" (release from Public Service Employees for Environmental Ethics, March 2003), http://www.pse.ca/Salmonpesticidebackgrounder032003.pdf; G.H. Rae, "Sea Louse Control in Scotland, Past and Present," *Pest Management Science* 58 (2002): 515–520; A. Ross, "UK Usage of Pesticides: Controls and Lessons to be Learned," in *Interactions between Aquaculture and the Environment,* ed. P. Oliver and E. Colleran (Dublin: An Taisce, 1990); M. Roth et al., "Current Practices in the Chemotherapeutic Control of Sea Lice Infestations: A Review," *Journal of Fish Diseases* 16 (1993): 1–26; M. Roth, "The Availability and Use of Chemotherapeutic Sea Lice Products," *Contributions to Zoology* 69 (2000): 109–118; R.A. Schnick, "The Impetus to Register New Therapeutants for Aquaculture," *Progressive Fish Culturist* 50 (1988): 190–196; R.A. Schnick, "Approval of Drugs and Chemicals for Use by the Aquaculture Industry," *Veterinary and Human Toxicology* 40 (Supplement) (1998): 9–17; P. Scott, "Therapy in Aquaculture," in *Aquaculture for Veterinarians,* ed. L. Brown (Oxford: Pergamon, 1993); "The Future for Sea Lice Control in Cultured Salmonids: A Review" (Perth: Scottish Wildlife and Countryside Link, 1992); "Bacterial Disease Control, Antibiotics and the Environment in Marine Finfish Culture: A Review" (Perth: Scottish Wildlife and Countryside Link, 1993); Statistics Norway, "Amount of Pharmaceuticals Sold for Use in Fish Farming (some substances): Agents Used against Endoparasites/Ektoparasites (1989-2001)" (report from Statistics Norway, 2003), http://www.ssb.no/english/subjects/10/05/nos_fiskeoppdrett_en/nos_d259_en/tab/4.2.html.

33. M. Roth, "The Availability and Use of Chemotherapeutic Sea Lice Products." *Contributions to Zoology*, 69 (1/2) (2000), http://

dpc.uba.uva.nl/ctz/vol69/nr01/a12.

34. "Final Report of the MPMMG Subgroup on Marine Fish Farming," Scottish Fisheries Working Paper No. 3 (Aberdeen: Scottish Office, 1992).
35. Information supplied by the Veterinary Medicines Directorate, February 17, 2004.
36. "Concern over Fish Farm Pesticides—Over 1000 Licences for Chemicals Classified as Pollutants," *Sunday Herald,* April 25, 2004, www.sundayherald.com/41587; "Scotland's Toxic Toilets Revealed—Filthy Five Named and Shamed," *Salmon Farm Monitor,* April 25, 2004, www.salmonfarmmonitor.org/pr250404.shtml.

Canthaxanthin

37. L.J. Forristal, "Is Something Fishy Going On?" *The World and I,* May 2000, http://www.worldandi.com/public/2000/may/fishy.html; "Vitamin Price Fixing Lawsuits," *The Rubins,* December 28, 2003, http://www.therubins.com/legal/vitamsuit.htm.
38. View Hoffmann-La Roche's "SalmoFan" at http://www.cbc.ca/consumers/market/files/food/salmon/colour.html.
39. E. Cha, "The 15 Colours of Salmon," *Wired* 12, no. 2 (February 2004), http://www.wired.com/wired/archive/12.02/start.html?pg=3; DSM Nutritional Products website is at http://www.roche-vitamins.com/home/what/what-gen/what-gen-carot/what-gen-carot-cantha.htm or http://www.roche-vitamins.com/home/what/what-hnh/what-hnh-prod/what-hnh-prod-caro.htm.
40. S. Andersen, "Salmon Color and the Consumer: Eye Appeal Is Buy Appeal" (report for International Institute of Fisheries Economics and Trade, 2000), http://oregonstate.edu/Dept/IIFET/2000/abstracts/andersons.html.
41. "Nerve Poison Found in Supermarket Salmon," *Independent on Sunday,* October 28, 1990.
42. "Final Report of the MPMMG Subgroup on Marine Fish Farming," Scottish Fisheries Working Paper No. 3 (Aberdeen: Scottish Office, 1992).
43. L.J. Forristal, "The Great Salmon Scam," *Insight on the News,* June 12, 2000, http://www.findarticles.com/cf_dls/m1571/22_16/62741745/ print.jhtml.
44. B. Daicker et al., "Canthaxanthin Retinopathy: An Investigation by Light and Electron Microscopy and Physiochemical Analysis," *Graefes Arch Clin Exper Ophthal* 225 (1987): 189–197; L.I. Lonn, "Canthaxanthin Retinopathy," *Arch Ophthalmol* 105 (1987): 1590–1591; C. Harnois et al., "Canthaxanthin Retinopathy: Anatomic and Functional Reversibility," *Arch Ophthalmol* 107 (1989): 538–540; G.B. Arden and F. Barker, "Canthaxanthin and the Eye: A Critical Ocular Toxicological Assessment," *J Toxicol Cut Ocular Toxicol* 10 (1991): 115–155; T.S. Chang et al., "Asymmetric Canthaxanthin Retinopathy," *Am J Ophthalmol* 119 (1995): 801–802, http://musclemonthly.com/articles/010115/010115-haycock-supplement-science.htm.
45. Letter from Hoffman-La Roche dated September 11, 1986, http://www.nutri.com/wn/thax.txt.

46. N. Craven, "Pink Poison," *Daily Mail,* December 24, 2002, http://www.jcaa.org/JCNL0302/Poison.htm.
47. Old Muckspreader, "Down on the Farm," *Private Eye,* July 24, 1987.
48. P.H. Rose et al, "Canthaxanthin Potential Tumorigenic and Toxic Effects in Prolonged Dietary Administration to Mice" (unpublished Report HLR 135/861058 submitted to WHO by Hoffmann-La Roche, Basle, 1987); W. Köpcke et al, "Canthaxanthin Deposition in the Retina: A Biostatistical Evaluation of 411 Cases Taking this Carotenoid for Medical or Cosmetical Purposes" (unpublished manuscript submitted by Hoffmann-La Roche, Basle, 1992), http://europa.eu.int/comm/food/fs/sc/oldcomm7/out10_en.html.
49. European Commission, "Opinion of the Scientific Committee on Animal Nutrition on the Use of Canthaxanthin in Feedingstuffs for Salmon and Trout, Laying Hens and Other Poultry" (report from the Health and Consumer Protection Directorate, Brussels, 2002), http://europa.eu.int/comm/food/fs/sc/scan/out81_en.pdf; D. Thompson, "Health Fears as Chemical Linked to Eye Defects Is Used to Dye Fish Pink," *Daily Mail,* January 25, 2002, http://list.zetnet.co.uk/ pipermail/seatrout-rev/2002-January/000061.html; D. Thompson, "Salmon Dye Can Damage Eyesight of Consumers—Watchdog Warns that Levels of Chemical Are Far Too High," *Daily Mail,* June 24, 2002, http://list.zetnet.co.uk/pipermail/seatrout-rev/2002-June/000180.html; European Commission, "Brighter Eyesight or Brighter Salmon?" (report from the Health and Consumer Protection Directorate, January 27, 2003), http://europa.eu.int/rapid/start/cgi/guesten.ksh?p_action.gettxt=gt&doc=IP/03/123|0|RAPID&lg=EN.
50. A. Osborn and J. Meikle, "Salmon Pink Becomes a Grey Area for EU," *The Guardian,* January 28, 2003, http://www.guardian.co.uk/food/Story/0,2763,883617,00.html.
51. G. Parker and J. Mason, "Grey Days ahead for EU Salmon," *Financial Times,* January 28, 2003, http://search.ft.com/search/ article.html?id=030128000513.
52. The Salmon Farm Protest Group, "Canthaxanthin Consultation Response to the Food Standards Agency," *Salmon Farm Monitor,* 2003, http://www.salmonfarmmonitor.org/sfpgreports.shtml.
53. European Commission, "Opinion of the Scientific Committee on Animal Nutrition on the Use of Canthaxanthin in Feedingstuffs for Salmon and Trout, Laying Hens and Other Polutry" (April 2002), http://europa.eu.int/comm/food/fs/sc/scan/out81_en.pdf.
54. Smith and Lowney, "The Colour Salmon Lawsuit" (2004), http://www.smithandlowney.com/salmon/information/.
55. L.J. Forristal, "The Great Salmon Scam."

Dichlorvos

56. "Dichlorvos (DDVP): A Hazardous Organophosphate," *Pesticide News*, September 1995, http://www.pan-uk.org/pestnews/actives/dichlorv.htm.
57. "Biological Monitoring of Workers Exposed to Organophosphorous Pesticides: A Guidance Note" (London: Health and Safety Executive, 1987).
58. International Programme on Chemical Safety, "Dichlorvos: Health and Safety Guide," IPCS Health and Safety Guide No. 18 (Geneva: World Health Organization, 1988), http://www.inchem.org/documents/ hsg/hsg/hsg018.htm; IPCS, "Dichlorvos," IPCS Environmental Health Criteria No. 79 (Geneva: WHO, 1989).
59. T. Needham, "The Ted Needham Column," *Fish Farmer*, Sept/Oct 1988, http://list.zetnet.co.uk/pipermail/seatrout-rev/2002-January/ 000059.html.
60. S. Poulter, "Cancer Scare over Fly Spray Used By Millions," *Daily Mail*, December 15, 2001, http://list.zetnet.co.uk/pipermail/seatrout-rev/2002-January/000058.html; UK Health and Safety Executive, "Ministers Act over a Range of Insecticides Containing Dichlorvos," April 19, 2002, http://www.hse.gov.uk/press/2002/e02076.htm.
61. R. Wootten et al., "Aspects of the Biology of the Parasitic Copepods on Farmed Atlantic Salmonids and Their Treatment," *Proceedings of the Royal Society of Edinburgh* 81, B (1982): 185–197.
62. G.H. Rae, "On the Trail of the Sea Lice," *Fish Farmer* 2 (1979): 22–25.
63. A. Ross and P. Horsmann, "The Use of Nuvan in the Salmon Farming Industry (report for the Marine Conservation Society, Ross-on-Wye, 1988).
64. UK House of Commons, *Agriculture Committee Inquiry into "Fish Farming in the UK" (1989-1990): Volume 2: Minutes of Evidence and Appendices*, 1990.
65. "Police Probe Fish Farm Poison Find," *The Press and Journal*, May 9, 1989.
66. "Loss of Nuvan Container Sparks New Call for Safety Measures," *West Highland Free Press*, October 27, 1989.
67 I.M. Davies, "Actual Amounts of Aquagard Used on Scottish Salmon Farms: Report to the Ministry and Agriculture Fisheries and Food and the Veterinary Medicines Directorate" (private and confidential report not available from the Scottish Executive, 1991).
68. J.G. McHenery et al., "Effects of Dichlorvos Exposure on the Acetylcholinesterase Levels of the Gills of the Mussel, Experimental and Field studies," Scottish Fisheries Working Paper 16/91 (Aberdeen: Scottish Office, 1991).
69. "Final Report of the MPMMG Subgroup on Marine Fish Farming," Scottish Fisheries Working Paper No. 3 (Aberdeen: Scottish Office, 1992); A. Pike, "Sea Lice: Major Pathogens of Farmed Atlantic Salmon," *Parasitology Today* 5, no. 9 (1989): 291–297; Graph of dichlorvos usage in Western Isles is available at http://www.w-isles.gov.uk/wies2-4.htm#aqua.
70. T. Needham, "Elimination of the Salmon Louse Using DDVP—Nogos 50"

(Unilever Research Summary Report, private and confidential, 1978).
71. "Ciba Geigy Admits Dichlorvos Is Not Recommended for Use of Salmon Farms," *New Scientist,* August 4, 1988.
72. UK House of Commons, *Agriculture Committee Inquiry into "Fish Farming in the UK" (1989-1990): Volume 2: Minutes of Evidence and Appendices,* 1990.
73. Water Research Centre, "Proposed Provisional Environmental Quality Standards for Dichlorvos in Water" (report for the Department of the Environment DoE 2249-M/2, 1991), http://www.fwr.org/ environs/dwi0119.htm.
74. OSPAR, "PARCOM Recommendation 94/6 on Best Environmental Practice for the Reduction of Inputs of Potentially Toxic Chemicals from Aquaculture Use" (report from Oslo and Paris Conventions for the Prevention of Marine Pollution, 16th joint meeting, 1994); Scottish Executive, "PARCOM Recommendation 94/6 on Best Environmental Practice (BEP) for the Reduction of Inputs of Potentially Toxic Chemicals from Aquaculture Use: United Kingdom Code of Best Environmental Practice," 2000.
75. M.W. Jones et al., "Reduced Sensitivity of the Salmon Louse to the Organophosphate Dichlorvos," *Journal of Fish Diseases* 15 (1992): 191–202; D.W. Wells et al., "Fate of Dichlorvos in Sea Water Following Treatment for Salmon Louse Infestation in Scottish Fish Farms," Scottish Fisheries Working Paper No. 13/90 (Aberdeen: Scottish Office, 1990).
76. M.W. Jones, "S Is for Salmosan—Still a Very Useful Sealice Treatment," *Fish Farming Today,* June 2003.
77. J.W. Treasurer, "Sea Lice Management Methods in Scotland," *Caligus,* December 1998, http://www.ecoserve.ie/projects/sealice/caligus5.html.
78. R. Edwards, "Flyspray Ban Urged As Cancer Fears Rise—Activists Say the Chemical in Flyspray is Deadly to Humans," *Sunday Herald,* January 20, 2002, http://www.sundayherald.co.uk/21665.
79. P.O. Brandal and E. Egidius, "Preliminary Report on Oral Treatment against Sea Lice with Neguvon," *Aquaculture* 10 (1977): 177–178; P.O. Brandal and E. Egidius, "Treatment of Salmon Lice with Neguvon: Description of Method and Equipment," *Aquaculture* 18 (1979): 183–188; O.B. Samuelson, "Degradation of Trichlorfon to Dichlorvos in Seawater: A Preliminary Report," *Aquaculture* 60 (1987): 161–164; K. Grave et al., "Utilisation of Dichlorvos and Trichlorfon in Salmonid Farming in Norway During 1981–1988," *Acta Veterinaera Scandica* 32 (1991): 1–7.
80. T. Horsberg, "Food Safety Aspects of Aquaculture Products in Norway" (release from Atlantic Institute for Market Studies, Halifax, NS, 2000), http://www.aims.ca/Aqua/horsberg.htm.
81. A.K. Pal, "Acute Toxicity of DDVP to Fish, Plankton and Worm," *Environment and Ecology* 1 (1983): 25–26; A.K. Pal and S.K. Konar, "Influence of the Organophosphorus Insecticide DDVP on Feeding, Survival, Growth and

Reproduction of Fish," *Environment and Ecology* 3 (1985): 398–402; E. Egidius and B. Moster, "Effect of Neguvon and Nuvan Treatment on Crabs, Lobster and Blue Mussel," *Aquaculture* 60 (1987): 165–168; N.S. Mattson et al., "Uptake and Elimination of Trichlorfon in Blue Mussel and European Oyster: Impact of Neguvon on Mollusc Farming," *Aquaculture* 71 (1988): 9–14; For a general review of scientific papers on dichlorvos see http://www.ecoserve.ie/projects/sealice/nuvan.html.

82. BC Ministry of Agriculture and Fisheries, "Nuvan, Sea Lice and Salmon Farming" (report from Aquaculture and Commercial Fisheries Branch, 1990).
83. R. Cusack and G. Johnson, "A Study of Dichlorvos: A Therapeutic Agent for Sea Lice," ERDA (Nova Scotia) Report No. 14, 1988.
84. R. Salte et al., "Fatal Acetylcholinesterase Inhibition in Salmonids Subjected to a Routine Organophosphate Treatment," *Aquaculture* 61 (1987): 173–179; J.P. Fraser et al., "Effects of a Cholinesterase Inhibitor on Salmonid Lens, a Possible Cause for the Increased Incidence of Cataract in Salmon," *Experimental Eye Research* 49 (1989): 293–98; T.E. Horsberg et al., "Organophosphate Poisoning of Atlantic Salmon in Connection with Treatment against Salmon Lice," *Acta. Vet. Scand.* 30 (1989): 385–390; P. Dobson and H.J. Schuurman, "Possible Causes of Cataract in Atlantic Salmon," *Experimental Eye Research* 50 (1990): 439–442; P.J. Fraser et al., "Nuvan and Cataracts in Atlantic Salmon," *Experimental Eye Research* 50 (1990): 443–447; D.W. Bruno et al., "The Use of Aquagard and the Prevalence of Cataracts Among Farmed Atlantic Salmon," Mariculture Committee CM 1991/F (Copenhagen: International Council for the Exploration of the Sea, 1991), 1–5; J.E. Dorfman Hecht et al., "Acute Effects of Nuvan on the Optical and Biochemical Properties of Cultured Atlantic Salmon Lenses," *In Vitro Toxicol.* 7 (1994): 339–349.
85. L.E. Burridge and K. Haya, "A Review of Di-n-butylphthalate in the Aquatic Environment: Concerns Regarding Its Use in Salmonid Aquaculture," *Journal of the World Aquaculture Society* 26 (1995): 1–13.
86. M. Roth et al., "Current Practices in the Chemotherapeutic Control of Sea Lice Infestations: A Review," *Journal of Fish Diseases* 16 (1993): 1–26.
87. Ciba Geigy, "Dichlorvos, (DDVP), A Report on Glass House Trials Done in the United Kingdom," Project Report 0-2763/1D, 1964; Ciba Geigy, "Effects of Dichlorvos on Human Cells in Tissue Culture," Temana-2 and Ciba-Geigy, Medisch Biologisch Laboratorium, Report No. 69, 1971; Ciba Geigy, "Report on Mutagenic Effect of Technical DDVP," Tierfarm AG, Report No. CP 974/58, Ciba Agriculture, Novartis, 1971.
88. Ciba Geigy, "NUVAN, Residues in Stored Rice and Cocoa Beans," CH. Project Report 11-11-65 (unpublished report, 1965); Ciba Geigy, "Dichlorvos, Determination of Dichlorvos in Barley after Application of NUVAN 7, trial silo Dintikon," CH, V 82, 1976; Ciba Geigy, "Dichlorvos, Gas Chromatographic

Residue Method for Plant Material, Meat, Cheese, Milk and Stored Cereals," REM-5-77 (unpublished report, 1977); Ciba Geigy, "Determination of Residues of Dichlorvos in Fat, Milk and Animal Tissues after Single Spray Application of NUVAN," 1986; Full reference list is at http://www.fao.org/waicent/faoinfo/agricult/agp/agpp/pesticid/JMPR/Download/93/dichlorv.pdf.

89. T.E. Horsberg and T. Hoy, "Residues of Dichlorvos in Atlantic Salmon after Delousing," *Journal of Agricultural Food Chemistry* 38 (1990): 1403–1406.
90. "Nerve Poison Found in Supermarket Salmon," *Independent on Sunday,* October 28, 1990; "Salmon Sales Fall after Nerve Poison Report," *Independent on Sunday,* November 4, 1990.
91. O. Tully and Y. McFadden, "Variation in Sensitivity of Sea Lice to Dichlorvos on Irish Salmon Farms in 1991–92," *Aquaculture Research* 31, no. 11 (2000): 849–854.
92. C. Kelleher et al., "Incidence of Leukemia, Lymphoma and Testicular Tumours in Western Ireland," *Journal of Epidemiology and Community Health* 52 (1998): 651–656.
93. "Cancers Blamed on Land Chemicals," BBC News, October 1, 1998, http://news.bbc.co.uk/1/hi/health/184215.stm; "Cancer Rise May Be Linked to Farm Chemicals," Reuters, October 16, 1998, http:// 131.104.232.9/agnet/1998/10-1998/ag-10-16-98-1.txt.
94. A.E. Czeizel, "Human Germinal Mutagenic Effects in Relation to Intentional and Accidental Exposure to Toxic Agents," *Environmental Health Perspectives* 104, suppl. 3 (1996): 615–617, http:// website.lineone.net/~mwarhurst/pesticides.html.
95. Statement from Dr. Andrew Grant of Novartis, December 19, 2001.
96. UK Department of Health, Committee on Mutagenicity, "Statement on Mutagenicity of Dichlorvos," July 2001; UK Health and Safety Executive, "Ministers Act over a Range of Insecticides Containing Dichlorvos," April 19, 2002, http://www.hse.gov.uk/press/2002/e02076.htm; "Insecticide Ban Amid Cancer Fears," BBC News, April 19, 2002, http:// news.bbc.co.uk/1/hi/uk/1939569.stm.
97. M.J. Ashwood-Smith et al., "Mutagenicity of Dichlorvos," *Nature* 240 (1972): 418–420; C.E. Voogd et al., "Mutagenic Action of Dichlorvos," *Mutation Research* 16 (1972): 413–416; B.J. Dean and D. Blair, "Dominant Lethal Assay in Female Mice after Oral Dosing with Dichlorvos or Exposure to Atmospheres Containing Dichlorvos," *Mutation Research* 40 (1976): 67–72; National Cancer Institute, "Bioassay of Dichlorvos for Possible Carcinogenicity" (Bethesda, MD: Carcinogen Bioassay and Program Resouces Branch, Carcinogenesis Program, National Cancer Institute, NIH, 1977); P. Perocco and A. Fini, "Damage by Dichlorvos of Human Lymphocyte DNA," *Tumori* 66 (1980): 425–430; R. Cusack and G. Johnson, "A Study of Dichlorvos: A Therapeutic Agent for Sea Lice," ERDA (Nova Scotia) Report No. 14 (1988); International Programme on Chemical Safety, "Dichlorvos: Health and Safety Guide," IPCS Health and Safety Guide No. 18 (Geneva: World Health Organization, 1988), http://www.inchem.org/

documents/hsg/hsg/hsg018.htm; IPCS, "Dichlorvos," IPCS Environmental Health Criteria No. 79 (Geneva: World Health Organization, 1989); National Toxicology Program, "NTP Technical Report on the Toxicology and Carcinogenesis Studies of Dichlorvos (CAS 62-73-7) in F344/N Rats and B6C3F1 Mice (Gavage Studies)," NIH Publication 89-2598 (Bethesda, MD: National Institutes of Health, 1989); O. Tully and D. Morrisey, "Concentrations of Dichlorvos in Beirtreach Bui Bay, Ireland," *Marine Pollution Bulletin* 20 (1989): 190–191; R. Cusack and G. Johnson, "A Study of Dichlorvos (Nuvan), a Therapeutic Agent for the Treatment of Salmonids Infected with Sea Lice," *Aquaculture* 90 (1990): 101–112; J.G. McHenery and C. Francis, "The Toxicity of Dichlorvos to Stage 4 *Homarus gamarus* Larvae," Scottish Fisheries Working Paper No. 8 (Aberdeen: Scottish Office, 1990); R.C.T. Raine et al., "Toxicity of Nuvan and Dichlorvos towards Marine Phytoplankton," *Botanica Marina* 33 (1990): 533–537; J.E. Thain et al., "The Toxicity of Dichlorvos to Some Marine Organisms," Marine Environment Quality Committee, E:18 (in mimeo), (Copenhagen: International Council for the Exploration of the Sea, 1990); J.G. McHenery et al., "Lethal and Sub-Lethal Effects of the Salmon Lousing Agent Dichlorvos on the Larvae of the Lobster and Herring," *Aquaculture* 98, no. 4 (1991): 331–347; J.G. McHenery et al., "Effects of Dichlorvos Exposure on the Acetylcholinesterase Levels of the Gills of the Mussel, Experimental and Field Studies," Scottish Fisheries Working Paper No. 16/91 (Aberdeen: Scottish Office, 1991); Anon, "Carcinogen Profile: Dichlorvos (DDVP)," *Prop 65 News* 6, no. 7 & 8 (July–August 1992), http://www.prop65news.com/pubs/p65news/issues/9207/920714.html; H.L. Bris et al., "Laboratory Study on the Effect of Dichlorvos on Two Commercial Bivalves," *Aquaculture* 138 (1995): 139–144; L.E. Burridge and K. Haya, "A Review of Di-n-butylphthalate in the Aquatic Environment: Concerns Regarding Its Use in Salmonid Aquaculture," *Journal of the World Aquaculture Society* 26 (1995): 1–13; J.G. McHenery et al., "Threshold Toxicity and Repeated Exposure Studies of Dichlorvos on the Larvae of the Common Lobster," *Aquatic Toxicology* 34 (1996): 237–251; D.J. Murrison et al., "Epiphyte Invertebrate Assemblages and Dichlorvos Usage at Salmon Farms," *Aquaculture* 159 (1997): 53–66; M.L. Kent, "Marine Netpen Farming Leads to Infections with Some Unusual Parasites," *International Journal for Parasitology* 30 (2000): 321–326; O. Tully and Y. McFadden, "Variation in Sensitivity of Sea Lice to Dichlorvos on Irish Salmon Farms in 1991–92"; United States Environmental Protection Agency dichlorvos links are at http://www.epa.gov/pesticides/op/ddvp.htm (see particularly the report of the Cancer Assessment Review Committee http://www.epa.gov/pesticides/op/ddvp/carcrep.pdf, the report on human health risks http://www.epa.gov/pesticides/op/ddvp/hedrisk.pdf, and the report on ecological effects http://www.epa.gov/pesticides/ op/ddvp/efedrisk.pdf)

98. F. Cameron, "Dichlorvos: Imminent Scottish Legislation Could Give Added Scope for Product Withdrawal," Intrafish, May 14, 2002,

http://www.intrafish.com/article.php?articleID=23199&s=1.

99. "Scotland's Toxic Toilets Revealed—Filthy Five Named and Shamed," *Salmon Farm Monitor,* April 25, 2004, www.salmonfarmmonitor.org/ pr250404.shtml;
100. "Fish Farm Chemical Scare: Fish Farm Workers Could Be at Considerable Risk from a Chemical Called Nuvan Used to Treat Salmon for Infestations of Sea Lice" (release from Friends of the Earth Scotland, July 18, 1988).
101. In a February 25, 2002, e-mail, Graeme Walker of the Health and Safety Executive wrote: "I regret that despite this I find myself unable to assist with most of your areas of enquiry either because: HSE personnel involved in the late 80s/90s have left/retired, taking with them knowledge of historic practice and inspection activity within the industry; or the information was either never held, or if it was, not recorded by HSE e.g. tonnages of dichlorvos used on salmon farms in Scotland...in respect of when and where HSE first became aware of the use of dichlorvos on fish farms, I'm afraid that this forms part of the corporate knowledge which has been lost over time."
102. "Dichlorvos" (York: Pesticides Safety Directorate, 1995).
103. Environmental Working Group, "The English Patients: Human Experiments and Pesticide Policy," July 1998, http://www.ewg.org/reports/english/englishpr.html.
104. J. Vidal, "Students Paid to Eat Organo-Phosphates," *The Guardian,* July 30, 1998, http://www.mad-cow.org/aug98_news.html#ddd; "Human Guinea Pigs Tested Pesticides," BBC News Online, July 30, 1998, http://news.bbc.co.uk/hi/english/health/newsid_142000/142335.stm.
105. N. Connolly, "Worker Sues Fish Farm over Testicular Cancer Link," *Sunday Business Post,* March 10, 2002.
106. F. Cook, "I Lost My Girlfriend and My Career and Almost Lost My Mind—Because of a Chemical Used in Fish Farms," *Mail on Sunday,* December 14, 2003, http://www.pan-uk.org/press/PIU81.htm.
107. For scientific references on OPs and BSE see http://www.purdeyenvironment.com/Web%20references.htm#OP's,%20pesticides.

Azamethiphos

108. "Chemical Treatments for Sea-Lice Infestation in Farmed Salmon" (release from Scottish Environment Protection Agency, June 11, 1997), http://www.sepa.org.uk/news/releases/1997/cypermethrin.htm.
109. A.N. Grant, "Medicines for Sea Lice," *Pest Management Science* 58 (2002): 521–527.
110. "Azamethiphos: Summary Report" (London: European Agency for the Evaluation of Medicinal Products, 1999), http://www.emea.eu.int/pdfs/vet/mrls/000195en.pdf.
111. L.E. Burridge et al., "The Lethality of Anti-Sea Lice Formulations Salmosan (Azamethiphos) and Excis (Cypermethrin) to Stage IV and Adult Lobsters during Repeated Short-Term Exposures," *Aquaculture* 182 (2000): 72–85.

112. M. Roth, "The Availability and Use of Chemotherapeutic Sea Lice Products," *Contributions to Zoology* 69 (2000): 109–118.
113. T. Horsberg, "Food Safety Aspects of Aquaculture Products in Norway" (release from Atlantic Institute for Market Studies, Halifax, NS, 2000), http://www.aims.ca/Aqua/horsberg.htm.
114. "Environmental Objectives for Norwegian Aquaculture: New Environmental Objectives for 1998–2000" (Trondheim, Norway: Directorate for Nature Management, 1999), http://193.217.72.207/ filer/pdf/Aquaculture.pdf.
115. R. Edwards, "New Fish Farm Pesticides to Flood Scottish Lochs," *New Scientist,* March 29, 1997.
116. "Risk Assessment of Azamethiphos," Policy No. 17 (Stirling: Scottish Environment Protection Agency, 1998) http://www.sepa.org.uk/pdf/policies/17.pdf.
117. "Scotland's Toxic Toilets Revealed—Filthy Five Named and Shamed," *Salmon Farm Monitor,* April 25, 2004, www.salmonfarmmonitor.org/pr250404.shtml.
118. J. Duffus, "An Environmental Impact Assessment with Regard to the Possible Use of Azamethiphos to Control Sea-Lice in Salmon" (confidential report prepared for Ciba-Geigy Agriculture, 1992).
119. S. Traenkle, "Salmosan. Product Licence Submission, V271, Ecochemistry and Ecotoxicity," Volume 1: Expert Summary Report (submitted by Ciba-Geigy Agriculture, September 1992).
120. "Azamethiphos: Summary Report," The European Agency for the Evaluation of Medicinal Products, Committee for Veterinary Medicinal Products, http://www.emea.eu.int/pdfs/vet/mrls/000195en.pdf.
121. "Bath Treatment of Atlantic Salmon with Azamethiphos" (document submitted by Hydro Seafood GSP to the Scottish Environment Protection Agency, February 9, 2000).
122. L.E. Burridge et al., "The Lethality of Salmosan (Azamethiphos) to American Lobster Larvae, Post-Larvae and Adults," *Ecotoxicology and Environmental Safety* 43 (1999): 165–169; P. Abgrall et al., "Sublethal Effects of Azamethiphos on Shelter Use by Juvenile Lobsters (*Homarus americanus*)," *Aquaculture* 181 (2000): 1–10.
123. L.E. Burridge et al., "The Lethality of Anti-Sea Lice Formulations Salmosan (Azamethiphos) and Excis (Cypermethrin) to Stage IV and Adult Lobsters during Repeated Short-Term Exposures," *Aquaculture* 182 (2000): 72–85.
124. R. Edwards, "Poison Linked to Fish Farms," *Sunday Herald,* February 6, 2000, http://www.sundayherald.com/6758.
125. Natural History Museum, "Environmental Impact of Sea-Lice Treatments," Phase 1 report. Department of the Environment contract EPG1/5/64 (confidential report for the Department of the Environment, the Veterinary Medicines Directorate, and the Veterinary Products Committee, 1997).
126. R. Edwards, "Big Catch: Fish Farming Is Flourishing at the Expense of Other Marine Life," *New Scientist,* April 17, 2002, http://list.zetnet.co.uk/

pipermail/seatrout-rev/2002-April/000164.html.

127. "Report of the Veterinary Products Committee to the Licensing Authority on Products with an Organophosphate as an Active Ingredient (Other Than Sheep Dips) (Addlestone: Veterinary Products Committee, 1999), http://www.vpc.gov.uk.
128. Advisory Committee on Pesticides, "New Genotoxicity Studies on Azamethiphos," ACP 7 (288/2001), http://www.pesticides.gov.uk /acp_home.asp# and "Human Health Review of Azamethiphos," ACP 63 (283/2001).
129. M.W. Jones, "S is for Salmosan—Still a Very Useful Sealice Treatment," *Fish Farming Today,* June 2003.

Cypermethrin

130. K. Boxaspen and J.C. Holm, "New Biocides Used against Sea Lice Compared to Organo-Phosphorus Compounds," in *Aquaculture and the Environment,* Special Publication No. 16, ed. N. De Pauw and J. Joyce (Ghent: European Aquaculture Society, 1991).
131. A.N. Grant, "Medicines for Sea Lice," *Pest Management Science* 58 (2002): 521–527.
132. G. Perger and D. Szadkowski, "Toxicology of Pyrethroids and Their Relevance for Human Health," *Annals of Agricultural and Environmental Medicine* 1 (1994): 11–17.
133. "Welsh Sheep Dip Monitoring Programme Summary Report" (Cardiff: Environment Agency, 1999).
134. S. Morgan, "Suit Filed against Fish Farmers Claiming Damage to Lobster by Pesticide," *Northern Aquaculture* 2, no. 11 (1996): 9; S. Morgan, "Guilty Plea to Using Cypermethrin," *Northern Aquaculture* 2 no. 3 (1996): 2; I. Muir, "Control of Sea Lice in Salmon Using Unlicensed Products," *The Veterinary Record,* February 22, 1997.
135. A. Moore and C.P. Waring, "The Effects of a Synthetic Pyrethroid on Some Aspects of Reproduction in Atlantic Salmon," *Aquatic Toxicology* 52 (2001): 1–12, http://www.cefas.co.uk/publications/files1001-1500/1062.htm.
136. "Study reference GP95033" (report from Grampian Pharmaceuticals Ltd. Research Division, 1995), 76; J.G. McHenery, "Review of Environmental Data Relating to the Use and Disposal of Cypermethrin Formulated as GPRDO1 (Excis) as a Bath Treatment for Salmon" (Inveresk Research International Project 384259, 1996); W. Ernst et al., "Dispersion and Toxicity to Non-Target Aquatic Organisms of Pesticides Used to Treat Sea Lice on Salmon in Net Pen Enclosures," *Marine Pollution Bulletin* 42, no. 6 (2001): 433–444, http://www.elsevier.com/gej-ng/10/32/47/34/30/25/abstract.html; B. Gowland et al., "Uptake and Effects of the Cypermethrin-Containing Sea Lice Treatment Excis in the Marine Mussel," *Environmental Pollution* 120, no. 3 (2002): 805–811; B. Gowland et al., "Cypermethrin Induces Glutathione S-Transferase Activity in the Shore Crab," *Marine Environmental Research* 54, no. 2 (2002): 169–177; K.J. Willis and N. Ling,

"Toxicity of the Aquaculture Pesticide Cypermethrin to Planktonic Marine Copepods," *Aquaculture Research* 35 (2004): 263–270.

137. P.J. Jacobsen and J.C. Holm, "Promising Test with New Compound against Sea Lice," *Norsk Fiskeoppdrett,* 16-18 (January 1990).
138. "Environmental Objectives for Norwegian Aquaculture: New Environmental Objectives for 1998–2000" (Trondheim, Norway: Directorate for Nature Management, 1999), http://193.217.72.207/ filer/pdf/Aquaculture.pdf.
139. Norwegian Directorate of Fisheries, "Key Figures from the Norwegian Aquaculture Industry, 2000" (Norway: Department of Agriculture, 2001), http://www.fiskeridir.no/english/pages/statistics/ keyfigures_aqua_00.pdf.
140. R.P. Hunter and N. Fraser, "Field Monitoring of the Effects of Cypermethrin as GPRDO1" (confidential report of Grampian Pharmaceuticals Ltd., Research Division. Ref GP95033, 1995).
141. W. Ernst et al., "Dispersion and Toxicity to Non-Target Aquatic Organisms of Pesticides Used to Treat Sea Lice..." *Marine Pollution Bulletin* 42(6):433-444, 2001, http://www.meramed.com/literature/ details.asp?ID=617.
142. SEPA successful prosecutions 1998–1999: Wadbister Offshore Ltd. Salmon Farm, Site 1, Laxfirth, Shetland (Illegal use of Deosan Deosect (cypermethrin) under the Control of Pollution Act (Section 30F3); Pled guilty at Lerwick Sheriff Court (April 1998) and fined £1,000). "Shetland Salmon Farm Fined Using Banned Chemical," Intrafish, April 30, 1998, http://www.intrafish.com/article.php?articleID=809.
143. A. Barnett, "'Illegal Poison' Used on Salmon: Chemical Treatment at Fish Farms Is Hazard to Health and Marine Life, Claims Ex-Employee," *The Observer,* April 30, 2000, http://www.observer.co.uk/uk_news/story/ 0,6903,215802,00.html.
144. "'Illegal' Use of Toxic Chemicals Revealed" (release from Friends of the Earth Scotland, April 30, 2000), http://www.foe-scotland.org.uk/press/ pr20000408.html.
145. A. Barnett, "'Illegal Poison' Used on Salmon." *The Observer,* April 30, 2000, http://observer.guardian.co.uk/uk_news/story/ 0,6903,215802,00.html.
146. "Illegal Chemical 'Used on Salmon,'" BBC News, July 14, 2000, http://news.bbc.co.uk/hi/english/uk/scotland/newsid_832000/832740.stm; "Question mark over Scottish 'quality' salmon," (release from Friends of the Earth Scotland, July 14, 2000), http://www.foe-scotland.org.uk/press/pr20000704.html.
147. "Cypermethrin Containers Washed up in Shetland" (release from Scottish Environment Protection Agency, August 10, 2001), http://www.sepa.org.uk/weeklybriefing/2001/aug/10082001.htm.
148. E. Barclay, "Fish Farmers in Chemicals Row," *Shetland Times,* August 10, 2001.
149. "Barricade Discovery Creates Storm in Shetland," Fisheries Information Service, August 14, 2001, http://www.fis.com/fis/worldnews/worldnews.asp?month year=&day=14&id=19730&l=e&country=&special=aquaculture.
150. T. Morton, "Tom Morton on Why For Him, the Salmon Is Off," *The Scotsman,*

August 15, 2001.

151. "Maine Uses Aquatic Formulation Cypermethrin," *Northern Aquaculture* 3, no. 1 (1997): 4.
152. "Risk Assessment of Cypermethrin," Policy No. 30 (Stirling: Scottish Environment Protection Agency, 1998), http://www.sepa.org.uk/pdf/guidance/fish_farm_manual/policy30.pdf.
153. B. Gowland et al., "Uptake and Effects of the Cypermethrin-Containing Sea Lice Treatment Excis in the Marine Mussel," *Environ. Pollution* 120(3):805-11, 2002, http://www.ncbi.nlm.nih.gov/entrez/query.fcgi?cmd=Retrieve&db=pubmed&dopt=Abstract&list_uids=12442804.
154. B. Gowland et al., "Cypermethrin Induces Glutathione S-Transferase Activity in the Shore Crab," *Marine Environ Res.* 54(2):169-77 August, 2002, http://www.ncbi.nlm.nih.gov/entrez/query.fcgi?cmd=Retrieve&db=PubMed&list_uids=12206409&dopt=Citation.
155. V. Zitko et al., "Toxicity of Permethrin, Decamethrin and Related Pyrethroids to Salmon and Lobster," *Bull. Environm. Contam. Toxicol.* 21 (1979): 338–343; D.W. McLeese et al., "Lethality of Permethrin, Cypermethrin and Fenvalerate to Salmon, Lobster and Shrimp," *Bull. Environm. Contam. Toxicol.* 25 (1980): 950–955; N.O. Crossland, "Aquatic Toxicology of Cypermethrin: Fate and Biological Effects in Pond Experiments," *Aquatic Toxicology* 2 (1982): 205–222; I.R. Hill, "Effects on Non-Target Organisms in Terrestrial and Aquatic Environments," in *The Pyrethroid Insecticides,* ed. J.P. Leahey (Philadelphia: Taylor and Francis, 1985); L.S. Mian and M.S. Mulla, "Effects of Pyrethroid Insecticides on Non-Target Invertebrates in Aquatic Ecosystems," *Agricultural Entomology* 9, no. 2 (1992): 73–98; U. Friberg-Jensen et al., "Effects of the Pyrethroid Insecticide, Cypermethrin, on a Freshwater Community Studied under Field Conditions: Direct and Indirect Effects on Abundance Measures of Organisms at Different Trophic Levels," *Aquatic Toxicology* 63, no. 4 (2003): 357–371.
156. V. Zitko et al., "Toxicity of Permethrin, Decamethrin and Related Pyrethroids to Salmon and Lobster"; D.W. McLeese et al., "Lethality of Permethrin, Cypermethrin and Fenvalerate to Salmon, Lobster and Shrimp"; L.E. Burridge and K. Haya, "The Use of a Fugacity Model to Assess the Risk of Pesticides to the Aquatic Environment on Prince Edward Island," *Advances in Environmental Science and Technology* 22 (1988): 193–201.
157. L.E. Burridge and K. Haya, "The Lethality of Pyrethrins to Larvae and Post-Larvae of the American Lobster," *Ecotoxicol. Environ. Saf.* 38, no. 2 (1997): 150–154; L.E. Burridge et al., "The Lethality of Anti-Sea Lice Formulations Salmosan (Azamethiphos) and Excis (Cypermethrin) to Stage IV and Adult Lobsters during Repeated Short-Term Exposures," *Aquaculture* 182 (2000): 72–85; L.E. Burridge et al., "The Lethality of the Cypermethrin Formulation Excis to Larval and Post-Larval Stages of the American Lobster, *Homarus americanus,*" *Aquaculture* 192

(2000): 37–47, http://www.scirus.com/search_simple/?query_1=cypermethrin +salmon&dsmem=on&offset=2; W. Ernst et al., "Dispersion and Toxicity to Non-Target Aquatic Organisms of Pesticides Used to Treat Sea Lice..."

158. R.R. Stephenson, "The Acute Toxicity of Cypermethrin (WL 43467) to the Freshwater Shrimp and Larvae of the Mayfly in Continuous Flow Tests" (confidential Shell Research Report No. TLGR-80.079, Sittingbourne, 1980).
159. "Research on the Dispersion of Sea Louse Pesticides in the Marine Environment," available on Department of Fisheries and Oceans Canada website http://www.mar.dfo-mpo.gc.ca/science/review/1996/Page/Page%20e.html.
160. W.E. Hogans, "Non-Target Organism Mortality Caused by Applications of Excis (Cypermethrin) Used in the Treatment of Sea Lice Parasites on Cultured Atlantic Salmon (report to the Department of Fisheries and Oceans, St. Andrews, 1997).
161. W. Ernst et al., "Dispersion and Toxicity to Non-Target Aquatic Organisms of Pesticides Used to Treat Sea Lice..." *Marine Pollution Bulletin* 42(6):433-444, 2001.
162. "Ecological Effects of Sea Lice Treatment Agents" (private and confidential report of Scottish Association for Marine Science, April 2001), http://list.zetnet.co.uk/pipermail/seatrout-rev/2002-April/000164.html.
163. K.J. Willis and N. Ling, "Toxicity of the Aquaculture Pesticide Cypermethrin to Planktonic Marine Copepods."
164. "The Occurrence of the Active Ingredients of Sea Lice Treatments in Sediments Adjacent to Marine Fish Farms" (Stirling: Scottish Environment Protection Agency, 2004), http://www.sepa.org.uk/aquaculture/projects/index.htm.
165. A. Moore and C.P. Waring, "The Effects of a Synthetic Pyrethroid on Some Aspects of Reproduction in Atlantic Salmon" *Aquatic Toxicology*, 52(1):1-12, March 2001.
166. "Question mark over Scottish 'quality' salmon," http://www.foe-scotland.org.uk/press/pr20000704.html.
167. Pesticides Action Network, "Cypermethrin: A Synthetic Pyrethroid," *Pesticides News* 30, December 1995.
168. Y. Shukla et al., "Carcinogenic and Cocarcinogenic Potential of Cypermethrin on Mouse Skin," *Cancer Letters* 182, no. 1 (2002): 33–41.
169. "Cypermethrin: Summary Report" (London: European Agency for the Evaluation of Medicinal Products, 1998).
170. G. Perger and D. Szadkowski, "Toxicology of Pyrethroids and Their Relevance for Human Health"; Pesticides Action Network, "Cypermethrin: A Synthetic Pyrethroid," *Ann Agri Environ Med* 1994 (1):11-17, http://www.aaem.pl/pdf/aaem9411.pdf.
171. "Question mark over Scottish 'quality' salmon," http://www.foe-scotland.org.uk/press/pr20000704.html.
172. I. Denholm et al., "Analysis and Management of Resistance to Chemotherapeutants in Salmon Lice," *Pest Management Science* 58 (2002): 528–536; A.N. Grant, "Medicines for Sea Lice," *Pest Management Science* 58 (2002):

521–527; S. Sevatdal and T.E. Horsberg, "The Identification of Pyrethroid Resistance in Sea Lice," *Norsk Fiskeoppdrett* 12 (2000): 34–35.

Teflubenzuron

173. B. Sauphanor and J.C. Bouvier, "Cross-Resistance between Benzyolureas and Benzoylhydrazines in the Codling Moth," *Pesticide Science* 45 (1995): 369–375.
174. A.N. Grant, "Medicines for Sea Lice," *Pest Management Science* 58 (2002): 521–527.
175. "Ectoban Will Help in Fight against Salmon Lice," *Fish Farming International* 8 (1995): 19; J.I. Erdal, "New Drug Treatment Hits Sea Lice When They Are Most Vulnerable," *Fish Farming International* 24, no. 2 (1997): 9; "Nutreco Ready with Its Curb on Sealice," *Fish Farming International* 24, no. 4 (1997): 3.
176. T. Horsberg, "Food Safety Aspects of Aquaculture Products in Norway" (release from Atlantic Institute for Market Studies, Halifax, NS, 2000), http://www.aims.ca/Aqua/horsberg.htm.
177. Norwegian Directorate of Fisheries, "Key Figures from the Norwegian Aquaculture Industry, 2000" (Norway: Department of Agriculture, 2001), http://www.fiskeridir.no/english/pages/statistics/ keyfigures_aqua_00.pdf.
178. T. Horsberg, "Food Safety Aspects of Aquaculture Products in Norway," Atlantic Institute for Market Studies, 2000, http://www.aims.ca/ Aqua/horsberg.htm.
179. "Target Salmon: Norwegian Environmental Group Is Warning Consumers Both at Home and Abroad against Farmed Salmon—Because It 'May Cause Cancer,' Intrafish, October 29, 1998, http://www.intrafish.com/article.php?articleID=2030.
180. T. Horsberg, "Food Safety Aspects of Aquaculture Products in Norway"; Norwegian Directorate of Fisheries, "Key Figures from the Norwegian Aquaculture Industry, 2000"; Statistics Norway, "Amount of Pharmaceuticals Sold for Use in Fish Farming: Agents Used against Endoparasites (1989–2001)" (report from Statistics Norway, 2003), http://www.ssb.no/english/subjects/10/05/nos_fiskeoppdrett_en/nos_d259_en/tab/4.2.html.
181. A.N. Grant, "Medicines for Sea Lice" *Pest Management Science*, 58:521-527, 2002.
182. T.E. Horsberg and T. Hoy, "Tissue Distribution of Diflubenzuron in Atlantic Salmon," *Acta Veterinaria Scandinavica* 32 (1991): 527–533.
183. "The Occurrence of the Active Ingredients of Sea Lice Treatments in Sediments Adjacent to Marine Fish Farms" (Stirling: Scottish Environment Protection Agency, 2004), http://www.sepa.org.uk/aquaculture/projects/index.htm.
184. "Teflubenzuron: Summary Report" (London: European Agency for the Evaluation of Medicinal Products, 1997).
185. L. Webster and P. Simpson, "Analysis of Teflubenzuron in Sediment and Biota by LC-MS," FRS Marine Laboratory Report No. 14 (Aberdeen: Scottish Marine Laboratory, 2001).
186. "The Toxicity of Sea Lice Chemotherapeutants to Non-Target Planktonic

Copepods" (Oban: Scottish Association for Marine Science, 2001).

187. M.N. Horst and A.N. Walker, "Biochemical Effects of Diflubenzuron on Chitin Synthesis in the Post-Smolt Blue Crab," *Journal of Crustacean Biology* 15 (1996): 401–408.
188. D. Ross and C. Holme, "Split on Use of Fish Farm Drugs: Shellfish Producers Call for Halt on Expansion Because of Unease over Increasing Use of Chemicals," *The Herald,* April 5, 2001; D. MacLeod, "Biotoxins Are Biggest Threat to the Industry," *Fish Farming Today,* August 2000.
189. Applications by Lighthouse of Scotland to the Scottish Environment Protection Agency, originally advertised in the *Edinburgh Gazette* May 2, 2003 (WPC/W/30736 Rubha Stillaig, Loch Fyne and WPC/W/30739 Arcastle, Loch Fyne).
190. "SEPA 'not tough enough on use of sea lice chemicals,'" *West Highland Free Press,* August 11, 2000.
191. Deep Trout, "Calicide: A Critique of Its Proposed Licence by SEPA as a Sea Lice Control Agent in Salmonid Aquaculture," 2000, http://ourworld.compuserve.com/homepages/BMLSS/Calcide.htm.
192. S. Naysmith, "Shellfish at risk from sea louse 'cure,'" *Sunday Herald,* November 26, 2000, http://www.sundayherald.com/12277.
193. Letter from Andy Rosie of SEPA to David Oakes, February 5, 2001.
194. "Long Term Environmental Monitoring of Teflubenzuron Used for the Treatment of Sea Lice in the Marine Environment," Addendum I and II to the Interim Report ARC-TFBZ-UK-5-98 (confidential report for Nutreco, 1998); Institute of Marine Research, "Tolerance of Juvenile Lobsters to a Feed Additive for Oral Treatment of Salmon Lice on Atlantic Salmon" (confidential report to Nutreco, 1995); "Environmental Risk Assessment of a Nutreco Insecticide" (confidential report to Nutreco, 1998); "Review of SEPA's Environmental Assessment of Teflubenzuron" (Water Research Centre report CO4650/2, 1999); G. Ritchie, "Long Term Environmental Monitoring of Teflubenzuron Used for the Treatment of Sea Lice in the Marine Environment" (confidential report to Nutreco, 1999); Institute of Aquaculture Stirling, "An ERA of the Use of Teflubenzuron to Control Ectoparasite Infestations on European Salmon Farms" (confidential report for Nutreco, 1998).
195. S. Naysmith, "'Dynamite' Report Reveals Fish Pollution," *Sunday Herald,* March 11, 2001, http://www.sundayherald.com/14244.
196. "Calicide (Teflubenzuron)—Authorization for Use as an In-Feed Sea Lice Treatment in Marine Cage Salmon Farms: Risk Assessment, EQS and Recommendations," Policy No. 29 (Stirling: Scottish Environment Protection Agency, 1999), http://www.sepa.org.uk/pdf/policies/29.pdf.
197. "Scotland's Toxic Toilets Revealed—Filthy Five Named and Shamed," *Salmon Farm Monitor,* April 25, 2004, www.salmonfarmmonitor.org/ pr250404.shtml.

198. "The Occurrence of the Active Ingredients of Sea Lice Treatments in Sediments Adjacent to Marine Fish Farms," http://www.sepa.org.uk/aquaculture/projects/index.htm.

Ivermectin

199. K.P.C. Hyland and S.J.R. Adams, "Ivermectin for Use in Fish," *Veterinary Record* 120 (1987): 539; I.M. Davies and G.K. Rodger, "A Review of the Use of Ivermectin as a Treatment for Sea Lice Infestation in Farmed Atlantic Salmon," *Aquaculture Research* 31, no. 11 (2000): 869–883.
200. L.E. Burridge et al., "The Lethality of the Cypermethrin Formulation Excis to Larval and Post-Larval Stages of the American Lobster, *Homarus americanus*," *Aquaculture* 192 (2000): 37–47, http://www.scirus.com/search_simple/?query_1=cypermethrin+salmon&dsmem=on&offset=2.
201. J.D. Pulliman and J.M. Preston, "Safety of Ivermectin in Target Animals," in *Ivermectin and Abamectin*, ed. W.C. Campbell (New York: Springer-Verlag, 1989).
202. R. Palmer et al., "Case Notes on Adverse Reactions Associated with Ivermectin Therapy of Atlantic Salmon," *Bulletin of the European Association of Fish Pathologists* 17, no. 2 (1996): 62–67; J.P.G. Toovey et al., "Ivermectin Inhibits Respiration in Isolated Rainbow Trout Gill Tissue," *Bulletin of the European Association of Fish Pathologists* 19, no. 4 (1999): 149–152.
203. R. Palmer et al., "Preliminary Trials on the Efficacy of Ivermectin against Parasitic Copepods of Atlantic Salmon," *Bulletin of the European Association of Fish Pathologists* 7, no. 2 (1987): 47–54; J. O'Halloran et al., "*Ergasilus labracis* on Atlantic Salmon," *Canadian Veterinary Journal* 33 (1992): 75; J. Kilmartin et al., "Investigations of the Toxicity of Ivermectin for Salmonids," *Bulletin of the European Association of Fish Pathologists* 17, no. 2 (1996): 58–61.
204. S.C. Johnson et al., "Toxicity and Pathological Effects of Orally Administered Ivermectin in Atlantic, Chinook and Coho Salmon and Steelhead Trout," *Diseases of Aquatic Organisms* 17 (1993): 107–112.
205. A. Morton, "What Is Wrong with Salmon Farming?" (report by Raincoast Research, 2003), http://www.raincoastresearch.org/salmon-farming.htm.
206. Q. Dodd, "BC Fish Loss Related to Lice Treatment," *Northern Aquaculture*, March 2000.
207. L.E. Burridge and K. Haya, "The Lethality of Ivermectin, a Potential Agent for Treatment of Salmonids against Sea Lice, to the Shrimp," *Aquaculture* 117 (1993): 9–14.
208. A. Morton, "What Is Wrong with Salmon Farming?"
209. "Ivermectin To Be Used on Scottish Salmon Farms," *New Scientist*, September 7, 1996, http://www.gn.apc.org/ecosystem/af968.htm#TARGET2.
210. B.A. Halley et al., "The Environmental Impact of the Use of Ivermectin: Environmental Effects and Fate," *Chemosphere* 18 (1989): 1543–1563.

211. J. Lichfield, "Ecologists Fight the Rise of Tough, Toxic Cowpats," *The Independent,* August 30, 2001, http://www.independent.co.uk/story.jsp?story=91337.
212. K.D. Black et al., "The Effects of Ivermectin, Used to Control Sea Lice on Caged Farmed Salmonids, on Infaunal Polychaetes," *ICES Journal of Marine Science* 54 (1997): 276–279; L.M. Collier and E.H. Pinn, "An Assessment of the Acute Impact of the Sea Lice Treatment Ivermectin on a Benthic Community," *Journal of Experimental Marine Biology and Ecology* 230, no. 1 (1998): 131–147; M. Costelloe et al., "Densities of Polychaetes in Sediments under a Salmon Farm Using Ivermectin," *Bulletin of the European Association of Fish Pathologists* 18, no. 1 (1998): 22–25; A. Cannavan et al., "Concentration of 22,23-dihydroavermectin B1a Detected in the Sediments at an Atlantic Salmon Farm Using Orally Administered Ivermectin to Control Sea-Lice Infestation," *Aquaculture* 182, no. 3–4 (2000): 229–240.
213. A. Grant and A. Briggs, "Toxicity of Ivermectin to Estuarine and Marine Invertebrates," *Marine Pollution Bulletin* 36, no. 7 (1998), 540–541; A. Grant and A. Briggs, "Use of Ivermectin in Marine Fish Farms: Some Concerns," *Marine Pollution Bulletin* 36, no. 8 (1998): 566–568; N. Nuttal, "Treatment for Farmed Salmon Highly Toxic," *The Times,* February 23, 1998, http://www.uea.ac.uk/~e130/ivermectin.html.
214. J.E. Thain et al., "Acute Toxicity of Ivermectin to the Lugworm, *Arenicola marina,*" *Aquaculture* 159 (1997): 47–52; I.M. Davies et al., "Environmental Risk of Ivermectin to Sediment-Dwelling Organisms," *Aquaculture* 163 (1998): 29–46; for further references on ivermectin use on salmon farms see http://www.ecoserve.ie/projects/sealice/ivermectin.html.
215. "Salmon Farms Could Be Taking a Lethal Toll on Marine Life," *New Scientist,* February 7, 1998.
216. I.M. Davies and G.K. Rodger, "A Review of the Use of Ivermectin as a Treatment for Sea Lice Infestation in Farmed Atlantic Salmon," *Aquaculture Research* 31, no. 11 (2000): 869–883.
217. "The Occurrence of the Active Ingredients of Sea Lice Treatments in Sediments Adjacent to Marine Fish Farms" (Stirling: Scottish Environment Protection Agency, 2004), http://www.sepa.org.uk/aquaculture/projects/index.htm.
218. T. Hoy et al., "The Deposition of Ivermectin in Atlantic Salmon," in *Chemotherapy in Aquaculture: From Theory to Reality,* ed. C. Michel and D.J. Alderman (Paris: Office International des Epizooties, 1992); D.G. Kennedy et al., "Determination of Ivermectin Residues in the Tissues of Atlantic Salmon Using HPLC with Fluorescence Detection," *Food Additives and Contaminants* 10, no. 5 (1993): 579–584; M. Roth, "Ivermectin Depuration in Atlantic," *Journal of Agricultural Food Chemistry* 41 (1993): 2434–2436.
219. J. Kay, "Analysis of Fish Flesh from Farmed Fish" (paper presented at the Royal Society of Chemistry conference "Fish Farming: Analysis All the Way to the Plate,"

October 19, 1995).

220. "Germans Shift Panic on British Food to Salmon," *The Times,* December 6, 1996.
221. "Salmon Boycott: Two British Supermarket Chains Are Refusing to Sell Salmon from Fish Farms Which Use a Pesticide Called Ivermectin to Kill Sea Lice," *New Scientist,* May 10, 1997.
222. "Scottish Salmon: Illegal Chemicals Found—Neuroinsecticide Found at Almost 10 Times 'Action Level'" (release from Friends of the Earth Scotland, September 9, 2001), http://www.foe-scotland.org.uk/press/pr20010907.html; "Illegal Chemicals Found in Scottish Salmon" (release from Friends of the Earth Scotland, September 22, 2000), http://www.foe-scotland.org.uk/press/pr20000911.html
223. "Unlicensed Sea Lice Drug Used in Fish Farms," *Irish Times,* December 22, 1990; J. O'Halloran and K. Coombs, "Treatment of Sea Lice on Atlantic Salmon with Ivermectin," *Canadian Veterinary Journal* 34 (1993): 505; S.C. Johnson and L. Margolis, "The Efficacy of Ivermectin for the Control of Sea Lice on Sea-Farmed Atlantic Salmon," *Diseases of Aquatic Organisms* 17 (1993): 101–106; S.C. Johnson and L. Margolis, "The Efficacy of Orally Administered Ivermectin for the Control of the Salmon Louse, Its Toxicological and Pathological Effects in Atlantic, Chinook and Coho Salmon and a Review of Its Tissue Residue Times and Its Effects on Non-Target Marine Organisms," Aquaculture Update No.64 (Nanaimo, Canada: Department of Fisheries and Oceans, Pacific Biological Station, 1994); A. Ross, "Conservationists against Ivermectin in Salmon Farming," *Pesticides News* 34 (December 1996), http://www.pan-uk.org/pestnews/Pn34/pn34p3.htm.
224. "Appeal lodged to aquaculture licence T12/85D by John Mulcahy of Save The Swilly," February 2003.
225. C. Clover, "Salmon in Shops To Be Tested for Pesticide," *Daily Telegraph,* July 8, 1991; C. Clover, "Pesticide Used Illegally on Salmon Farm," *Daily Telegraph,* June 29, 1992.
226. "Scottish Salmon Farmers Operating outside the Law" (release from Friends of the Earth Scotland, July 8, 1991).
227. "Fish Farm Escapes Lightly over Unlicensed Pesticide," *ENDSReport,* February 1993, http://www.endsreport.com/.
228. "Veil Cast over Salmon Farms Using Ivermectin: SSQC Mark Withdrawn from Offending Farms for Six Months," *Scottish Fish Farmer,* August 1995.
229. R. McNeil, "Rattling a Skeleton in the Salmon Industry Cupboard," *Shetland Times,* July 21, 1995, http://list.zetnet.co.uk/pipermail/ seatrout-rev/2002-January/000060.html.
230. G.H. Rae, "Guidelines for the Use of Ivermectin Premix for Pigs to Treat Farmed Salmon for Sea Lice" (Perth: Scottish Salmon Growers Association, 1996); J.G. McHenery, "Assessment of the Potential Environmental Impacts of the Use of Ivermectin as an In Feed Medication for Salmon" (private and confidential Inveresk Research International report for the Scottish Salmon Growers

Association, 1996).
231. Letter from Laurie MacBride of the Georgia Strait Alliance, July 21, 2001, http://www.georgiastrait.org/Articles2001/ButeCSRDJuly.php.
232. R. Edwards, "Salmon Farmers Win Licence To Kill," *New Scientist,* September 7, 1996, http://www.gn.apc.org/ecosystem/af968.htm#TARGET2.
233. "Ivermectin Discharge Consent Application Called In" (release from the Scottish Executive, February 8, 1999), http://www.scotland.gov.uk/news/releas99_2/pr0275.htm.
234. C. Watson, "Fish Farms Face Public Inquiry on Bid to Use Toxic Insecticide," *The Herald,* July 12, 1999; C. Watson, "Fish Farm Withdraws Insecticide Request," *The Herald,* August 25, 1999.
235. A. Barnett, "'Illegal Poison' Used on Salmon: Chemical Treatment at Fish Farms Is Hazard to Health and Marine Life, Claims Ex-Employee," *The Observer,* April 30, 2000, http://www.observer.co.uk/uk_news/story/ 0,6903,215802,00.html.
236. "'Illegal' Use of Toxic Chemicals Revealed" (release from Friends of the Earth Scotland, April 30, 2000), http://www.foe-scotland.org.uk/press/ pr20000408.html.
237. "Question mark over Scottish 'quality' salmon," (release from Friends of the Earth Scotland, July 14, 2000), http://www.foe-scotland.org.uk/ press/pr20000704.html; "Scottish Salmon Company Stripped of 'Quality Mark'" (release from Friends of the Earth Scotland, July 19, 2000), http://www.foe-scotland.org.uk/press/pr20000705.html.
238. "Setterness Salmon Ltd, Lerwick Sheriff Court" (release from Scottish Environment Protection Agency, January 23, 2002), http://www.sepa.org.uk/news/releases/2002/pr020.html; "Scottish Salmon Farmers in Court for 'Illegal' Chemical Use" (release from Friends of the Earth Scotland, January 23, 2002), "£6,000 Fine for Scottish Salmon Farmers for Illegal Chemical" (release from Friends of the Earth Scotland, January 23, 2002), http://www.foe-scotland.org.uk/press/pr20020108.html; "Shetland Fish Farm Caught Using Illegal Chemicals Withdraws from Quality Mark" (release from Friends of the Earth Scotland, January 29, 2002), http://www.foe-scotland.org.uk/press/pr20020111.html.
239. H. Marter, "Message Had Not Got Through," *Shetland Seafood News,* February 2001, http://www.fishing-news.co.uk/headlines/feb02/ sepa05.htm.

Emamectin benzoate

240. You can see the safety data sheet for Proclaim at http://www.yescotton.com/msds/affirm-msds.pdf.
241. A.N. Grant, "Medicines for Sea Lice," *Pest Management Science* 58 (2002): 521–527.
242. T.C. Telfer, "Marine Environmental Effects Monitoring and Dispersion Study of the Anti-Sea Lice Drug SCH5884 in Loch Duich under Commercial Use Conditions" (Schering Plough Animal Health Study No. 1090N-60-V97-357,

1998); W.J. Roy et al., "Tolerance of Atlantic Salmon and Rainbow Trout to Emamectin Benzoate, a New Orally Administered Treatment for Sea Lice," *Aquaculture* 184 (2000): 19–29; J. Stone et al., "Field Trials to Evaluate the Efficacy of Emamectin Benzoate in the Control of Sea Lice Infestations in Atlantic Salmon," *Aquaculture* 186 (2000): 205–219; R. Armstrong et al., "A Field Efficacy Evaluation of Emamectin Benzoate for the Control of Sea Lice on Atlantic Salmon," *Canadian Veterinarian Journal* 41 (2000): 607–612.

243. "New Toxic Chemical Is No 'Magic Bullet' (release from Friends of the Earth Scotland, January 17, 2001), http://www.foe-scotland.org.uk/press/pr20010102.html.
244. W.J. Roy et al., "Tolerance of Atlantic Salmon and Rainbow Trout to Emamectin Benzoate..."; J. Stone et al., "Safety and Efficacy of Emamectin Benzoate Administered In-Feed to Atlantic Salmon Smolts in Freshwater as a Preventative Treatment against Infestations of Sea Lice," *Aquaculture* 210: 21–34, 2002.
245. J. Stone et al., "Safety and Efficacy of Emamectin Benzoate Administered In-Feed to Atlantic Salmon Smolts in Freshwater..." *Aquaculture*, 210:21-34, 2002.
246. W.J. Roy et al., "Tolerance of Atlantic Salmon and Rainbow Trout to Emamectin Benzoate..." *Aquaculture*, 184:19-29.
247. "Emamectin Benzoate Use in Marine Fish Farms: An Environmental Risk Assessment," SEPA Board Paper 65 (Stirling: Scottish Environment Protection Agency, 1999), http://www.sepa.org.uk/pdf/aquaculture/policies/emamectin_benzoate.pdf.
248. "The Toxicity of Sea Lice Chemotherapeutants to Non-Target Planktonic Copepods" (Oban: Scottish Association for Marine Science, 2001).
249. K.J. Willis and N. Ling, "The Toxicity of Emamectin Benzoate, an Aquaculture Pesticide, to Planktonic Marine Copepods," *Aquaculture* 221 (2003): 289–297
250. S.L. Waddy et al., "Emamectin Induces Moulting in the American Lobster," *Canadian Journal of Fisheries and Aquatic Sciences* 59, no. 7 (2002): 1096–1099.
251. A.C. Chuckwudebe et al., "Toxicity of Emamectin Benzoate to Mallard Duck and Northern Bobwhite Quail," *Environ. Toxicol. Chem.* 17, no. 6 (1996): 1118–1123.
252. M. Mushtaq et al., "Immobility of Emamectin Benzoate in Soils," *Journal of Agriculture and Food Chemistry* 44 (1996): 940–944; A.C. Chuckwudebe et al., "Uptake of Emamectin Benzoate Residues from Soil by Rotational Crops," *Journal of Agricultural and Food Chemistry* 44, no. 12 (1996): 4015–4121; L.S. Crouch and W.F. Feely, "Fate of Emamectin Benzoate in Head Lettuce," *Journal of Agricultural and Food Chemistry* 45, no. 12 (1995): 3075–3087; W.F. Feeley and L.S. Crouch, "Fate of Emamectin Benzoate in Cabbage, 2. Unextractable Residues," *Journal of Agricultural and Food Chemistry* 45, no. 7 (1997): 2758–2762; M.R. Hicks, "Determination of Emamectin in Freshwater and Seawater at Picogram-per-Milliliter Levels by Liquid Chromatography with Fluorescence Detection," *Journal of AOAC International* 80, no. 5 (1997): 1098–1103; J.S. O'Grodnick et al., "Aged

Soil Column Leaching of Emamectin Benzoate," *Journal of Agricultural and Food Chemistry* 44, no. 12 (1998): 2044–2048.

253. K.J. Willis and N. Ling, "The Toxicity of Emamectin Benzoate, an Aquaculture Pesticide, to Planktonic Marine Copepods."
254. "The Occurrence of the Active Ingredients of Sea Lice Treatments in Sediments Adjacent to Marine Fish Farms" (Stirling: Scottish Environment Protection Agency, 2004), http://www.sepa.org.uk/aquaculture/projects/index.htm.
255. "Maine Regulators Asked to Re-evaluate Salmon Farming Rules after Escape, Residue Findings," Intrafish, February 16, 2004, http://www.intrafish.com/article.php?articleID=42077.
256. T.C. Telfer, "Marine Environmental Effects Monitoring and Dispersion Study of the Anti-Sea Lice Drug SCH5884..." Shering Plough Animal Health Study No. 1090N-60-V97-357, 1998.
257. J.G. McHenery and C.M. Mackie, "Revised Expert Report on the Potential Environmental Impacts of Emamectin Benzoate, Formulated as Slice, for Salmonids," Cordah Report No. SCH001R5 (confidential report to Schering Plough Animal Health, 1999); T. Nickell, "The Effects of Emamectin Benzoate on Infaunal Polychaetes," Final Report DML Project 20898 for Schering Plough Animal Health (Dunstaffnage Marine Laboratory Internal Report 226, 2002).
258. Schering Plough's website for background on Slice is http://www.myfishpharm.com/aquahome.html; http://www.spaquaculture.com/default.aspx?pageid=545.
259. "Schering Plough Signs Consent Decree with FDA—Agrees To Pay $500 Million" (release from US Food and Drug Administration, May 17, 2002), http://www.fda.gov/bbs/topics/NEWS/2002/NEW00809.html; "US Drug Manufacturer Fined $500 Million for Violating Good Manufacturing Practices Regulation," http://www.rainbowinvestigations.com/story.cfm?rainbowID=105.
260. M. Gordon, "Schering-Plough Fined $1M," *Daily Record,* October 9, 2003.
261. "Ill Salmon Treated with Unapproved Drug," *Globe and Mail,* June 14, 2002, http://www.iatp.org/foodsec/News/news.cfm?News_ID=1827.
262. "Salmon Farmers Sidestep Drug Ban: Millions of Salmon Routinely Treated with Unapproved Anti-Parasite Measures That Could Be an Environmental Danger," *Ottawa Citizen,* October 3, 2000.
263. "Assessment Report of the Canadian Food Inspection Agency Activities Related to the Safety of Aquaculture Products" (report from Health Canada, June 2001), http://www.hc-sc.gc.ca/food-aliment/fsa-esa/aquacult/e_aq-e08.htm.
264. "Aquaculture Chemicals: BC Government Ignores Their Own Law—The Unpermitted Use of Powerful Pesticides to Treat Sea Lice Violates Provincial Pesticide Control Act" (release from Public Service Employees for Environmental Ethics, March 13, 2003), http://www.pse.ca.
265. "Pesticide Use in Aquaculture" (release from Public Service Employees for

Environmental Ethics, March 2003), http://www.pse.ca/Salmonpesticidebackgrounder032003.pdf.

266. "Group Blows Whistle on Fish Farm Pesticide," CBC-TV report, March 7, 2003, http://vancouver.cbc.ca/regional/servlet/ View?filename=bc_slice20030307.
267. "SEPA to Speed Sea Lice Treatment Process" (release from Scottish Environment Protection Agency, February 19, 2002), http:// www.sepa.org.uk/news/releases/2002/pr026.html.
268. "Emamectin Benzoate Use in Marine Fish Farms: An Environmental Risk Assessment," SEPA Board Paper 65.
269. F. Cameron, "'Off-Label' Use of Slice Condemned on All Sides," Intrafish, August 8, 2003, http:// www.intrafish.com/ articlea.php?articleID=37087.
270. J.M. Clark et al., "Resistance to Avermectins: Extent, Mechanisms, and Management Implications," *Ann. Rev. Entomol.* 40 (1995): 1–30.
271. "A Sea Lice 'Silver Bullet'?" *Salmon Farm Monitor,* May 2003, http://www.salmonfarmmonitor.org/intlnewsmay2003.shtml#item7.

TBT

272. Interview with BBC Radio 4's "Face the Facts" in 1985.
273. A.W. Berry, "Tributyltin Cage Net Antifouling and Its Effect on Shellfish Farming and the Marine Environment: A Crime against Nature and the Shellfish Farming Industry" (Argyll: Knapdale Seafarms Ltd, 1992).
274. M.L.H. Thomas, "Experiments in the Control of Shipworm Using Tributyltin Oxide," Technical Report No. 21 (Ottawa: Fisheries Research Board of Canada, 1967).
275. B.S. Smith, "Tributyltin Compounds Induce Male Characteristics on Female Mud Snails," *Journal of Applied Toxicology* 1 (1981): 141–144; J.E. Thain, "The Acute Toxicity of Bis (Tributyltin) Oxide to the Adults and Larvae of Some Marine Organisms," ICES CM 1983 E:13 (Copenhagen: International Council for the Exploration of the Sea, 1983); A.R. Beaumont et al., "High Mortality of the Larvae of the Common Mussel at Low Concentrations of Tributyltin," *Marine Pollution Bulletin* 15 (1984): 402–405.
276. I.M. Davies et al., "Accumulation of Tin and Tributyltin from Anti-Fouling Paint by Cultivated Scallops and Pacific Oysters," *Aquaculture* 55 (1986): 103–114.
277. J. Blythman, "Salmon Farmers in for a Grilling," *Sunday Herald,* March 11, 2001, http://www.sundayherald.com/14198.
278. "Tin from Paint Is Found in Salmon Flesh," *New Scientist,* January 22, 1987.
279. J.W. Short and F.P. Thrower, "Accumulation of Butyltins in Muscle Tissue of Chinook Salmon Reared in Sea Pens Treated with Tributyltin," in proceedings of Oceans 86 (conference held in Washington DC, September 23–25, 1986): 1117–1181.
280. A. Duff, "More Trouble with TBT," *Marine Pollution Bulletin* 18, no. 2 (1987): 28.

281. I.M. Davies and J.C. McKie, "Accumulation of Total Tin and Tributyltin in Muscle Tissue of Farmed Atlantic Salmon," *Marine Pollution Bulletin* 18, no. 7 (1987): 405–407.
282. "Tributyltin Should Be Banned without Delay" (release from Friends of the Earth, February 6, 1987).
283. "Growing Outcry over Use of 'Lethal' Chemical in Fish Farm Operations," *West Highland Free Press*, February 20, 1987.
284. B. Wilson, "Government Bans Marine Use of Toxic Paints," *Glasgow Herald*, February 25, 1987.
285. I.M. Davies et al., "Effects of Tributyltin Compounds from Antifoulants on Pacific Oysters in Scottish Sea Lochs," *Aquaculture* 74 (1988): 319–330.
286. "Final Report of the MPMMG Subgroup on Marine Fish Farming," Scottish Fisheries Working Paper No. 3 (Aberdeen: Scottish Office, 1992).
287. R.C.T. Raine et al., "Toxicity of Nuvan and Dichlorvos towards Marine Phytoplankton," *Botanica Marina* 33 (1990): 533–537; R. Edwards, "Poison Linked to Fish Farms," *Sunday Herald*, February 6, 2000, http://www.sundayherald.com/6758; D. Staniford, "A Big Fish in a Small Pond: The Global Environmental and Public Health Threat of Sea Cage Fish Farming" (paper presented at a conference in Chile organized by Terram, June 5–6, 2002), http://www.watershed-watch.org/ww/publications/sf/BigFishSmallPond(Chile).pdf.
288. "Stockpiling of TBT-Based Paints," *West Highland Free Press*, April 3, 1987.
289. "The TBT Bonanza," *Shetland Times*, May 22, 1987.
290. "TBT Man Fined £600," *Shetland Times*, October 9, 1987.
291. J.M. Pirie and J.C. McKie, "Tributyltin Dumping at Skerries, Shetland Isles: Report of a Second Survey of Effects of TBT on Dogwhelks and Limpets," Scottish Fisheries Working Paper No. 7 (Aberdeen: Scottish Office, 1989).
292. I.M. Davies et al., "Effects of TBT in Western Coastal Waters" (private and confidential final report to DETR, contract PECD CW0691: Fisheries Research Services Report No. 5/98, not available from the Scottish Executive, 1998).
293. "Welcome, If A Bit Late," *Glasgow Herald*, February 25, 1987.
294. "The TBT Controversy: Copper Seen as Safer than Tin," *Fish Farming International*, December 1987.
295. B. Rygg, "Effects of Sediment Copper on Benthic Fauna," *Marine Ecology Progress Series* 25 (1985): 83–89; A. Tiltnes and A.M. Skullerud, "Marine Pollution Caused by Antifouling Agents" (SFT Report No. 78, Norway, 1987); K.A. Moe and G.M. Skeie, "Copper from Antifouling Agents" (SFT Report No. 87, Norway, 1988); O. Tryland and J. Oxnevad, "Material Flow Analysis of Copper: Assessment of Alternatives" (SFT Report No. 92, Norway, 1992).
296. A. Morton, "What Is Wrong with Salmon Farming?" (report by Raincoast Research, 2003), http://www.raincoastresearch.org/salmon-farming.htm.

297. "Environmental Objectives for Norwegian Aquaculture: New Environmental Objectives for 1998–2000" (Trondheim, Norway: Directorate for Nature Management, 1999), http://193.217.72.207/ filer/pdf/Aquaculture.pdf.
298. O. Hjellestad, "Proposal to Ban Copper Emissions," Fisheries Information Service, July 13, 2001, http://www.fis.com.
299. "Appeal lodged to aquaculture licence T12/85D by John Mulcahy of Save The Swilly," February 2003.
300. "Environmental Objectives for Norwegian Aquaculture: New Environmental Objectives for 1998–2000."
301 "Norwegian EPA Calls for Fish Farm Toxin Ban," *ENDSReport,* July 2001, http://www.environmentdaily.com/articles/ index.cfm?action=article&ref=10305&searchtext=fish&searchtype=All.
302. A.G. Lewis and A. Metaxas, "Concentrations of Total Dissolved Copper in and near a Copper-Treated Salmon Net Pen," *Aquaculture* 99 (1991): 269–276.
303. T.H. Johnsen et al., "Copper Loads in Connection with Net Washing and Impregnation of Fish Pens" (Norway: NIVA, 1996).
304. A. Rosie, "Legislation and the Control of Caged Fish Farms by River Purification Boards" (paper presented at the Royal Society of Chemistry conference "Fish Farming: Analysis All the Way to the Plate," October 19, 1995).
305. "An Assessment of Sediment Copper and Zinc Concentrations at Marine Caged Fish Farms in SEPA West region (SEPA West, East Kilbride)" (unpublished report by Scottish Environment Protection Agency, 1998).
306. D.J. Morrisey et al., "Predicting Impacts and Recovery of Marine Sites in Stewart Island, New Zealand, from the Findlay-Watling Model," *Aquaculture* 185 (2000): 257–271.
307. D. Staniford, "Closing the Net on Sea Cage Fish Farming" (paper presented at Charting the Best Course: The Future of Mariculture in Australia's Marine Environment, August 27, 2003), http://www.salmonfarmmonitor.org/indreports.shtml.
308. "National Residue Survey Results for 2001–2002" (Australian Department of Agriculture Fisheries and Forestry, 2003), http://www.affa.gov.au/content/ output.cfm?ObjectID=D2C48F86-BA1A-11A1-A2200060B0A05746& contType=outputs.

Malachite Green

309. N.C. Nelson, "A Review of the Literature on the Use of Malachite Green in Fisheries" (Springfield: National Technical Information Service, 1974).
310. D.J. Alderman, "Malachite Green: A Review," *Journal of Fish Diseases* 8 (1985): 289–298. N. Gerudo et al., "Pathological Effects of Repeated Doses of Malachite Green: A Preliminary Study," *Journal of Fish Diseases* 14 (1991): 521–532; D.J. Alderman, "Malachite Green and Alternatives as Therapeutic Agents," *European*

Aquaculture Society Special Publication 16 (1992): 235–244; D.J. Alderman, "Chemicals in the Hatchery," *Fish Farmer,* March/April 1997.

311. "Contaminated Chilean Salmon Impounded in Europe," *Salmon Farm Monitor,* August 2003, http://www.salmonfarmmonitor.org/intlnewsaugust 2003.shtml#item1; "Malachite Green Contamination in Chilean Salmon," *Salmon Farm Monitor,* September 2003, http://www.salmonfarmmonitor.org/ intlnewsseptember2003.shtml#item1.
312. "Nutreco Fined for Illegal Use of Malachite Green," *Salmon Farm Monitor,* September 2003, http://www.salmonfarmmonitor.org/intlnewsseptember 2003.shtml#item2; "Nutreco Caught in Malachite Scandal," *Salmon Farm Monitor,* October 2003, http://www.salmonfarmmonitor.org/intlnewsoctober2003. shtml#item4.
313. S. Carrell, "'Mutant' Chemicals Used in Fish Farms: Fish Farm Dye Link to Animal Mutation," *The Independent on Sunday,* March 25, 2001, http://list.zetnet.co.uk/pipermail/seatrout-rev/2002-June/000181.html; J. Reynolds, "Contaminated Salmon on Sale to Public," *The Scotsman,* August 5, 2003, http://www.news.scotsman.com/ topics.cfm?id=844672003&tid=15.
314. R. Edwards, "Europe Threat to Ban Toxic Salmon," *Sunday Herald,* December 15, 2002, http://www.sundayherald.com/29991; V. Fletcher, "EU Ban Looms for Our Farmed Salmon," *London Evening Standard,* December 16, 2002, http://persweb.direct.ca/kbryski/farmed-cancer.html; "Final Report on a Mission Carried out in the UK (16–26 April) in Order to Evaluate the Control of Residues in Live Animals and Animal Products" (European Commission, Health and Consumer Protection Directorate, October 29, 2002), http://europa.eu.int/comm/ food/fs/inspections/vi/reports/united_kingdom/vi_rep_unik_8626-2002_en.pdf.
315. T. Horsberg, "Food Safety Aspects of Aquaculture Products in Norway" (release from Atlantic Institute for Market Studies, Halifax, NS, 2000), http://www.aims.ca/Aqua/horsberg.htm.
316. "Illegal Medicine Used in Norwegian Marine Farming," Intrafish, August 3, 2001, http://www.intrafish.com/articlea.php?articleID=14738; "Professor Looking for Total Ban of Malachite Drug," Intrafish, August 6, 2001, http://www.intrafish.com/ articlea.php?articleID=14758.
317. R. Roberts, "The Lingering Legacy of Malachite Green," *Aquaculture Magazine,* March/April 2003, http://www.aquaculturemag.com/siteenglish/printed/ archives/issues03/203a.html.
318. "COM Statement on Malachite Green and Leucomalachite Green," (report of the Committee on Mutagenicity, UK Department of Health, 1999), http://www.dh.gov.uk; "Residues of Malachite Green in Farmed Fish: Statement by the Committee on Toxicity of Chemicals in Food, Consumer Products and the Environment on Surveillance for Malachite Green and Leucomalachite Green in Farmed Fish (UK Department of Health, 1999).

319. T.P. Glagoleva and E.M. Malikova, "The Effect of Malachite Green on the Blood Composition of Young Baltic Salmon," *Rybnoe Khozyaistvo* 45 (1968): 15–18; T.D. Bills and J.B. Hunn, "Changes in the Blood Chemistry of Coho Salmon Exposed to Malachite Green," *Progressive Fish Culturist* 38 (1976): 214–216; W.S. Fisher et al., "Toxicity of Malachite Green to Cultured American Lobster Larvae," *Aquaculture* 8 (1976): 151–156; F.P. Meyer and T.A. Jorgenson, "Tetralogical and Other Effects of Malachite Green on Development of Rainbow Trout and Rabbits," *Transactions of the American Fisheries Society* 112 (1983): 818–824; S. Clemmensen et al., "Toxicological Studies on Malachite Green: A Triphenylmethane Dye," *Archives of Toxicology* 56 (1984): 43–45; C. Fernandes et al., "Enhancing Effect of Malachite Green on the Development of Hepatic Pre-Neoplastic Lesions Induced by N-nitrosodiethylamine in Rats," *Carcinogenesis* 12 (1991): 839–845; S.J. Culp and F.A. Beland, "Malachite Green: A Toxicological Review," *J. Am. Coll. Toxicol.* 15 (1996): 219–238; S.J. Culp et al., (1997) Metabolic Changes Occurring in Mice and Rats Fed Leucomalachite Green—A Reduced Derivative of the Antifungal Agent Malachite Green," (from the proceedings of the 88th meeting of the American Society for Cancer Research in San Diego, 1996); V. Fessard et al., "Mutagenicity of Malachite Green and Leucomalachite Green in In Vitro Tests," *Journal of Applied Toxicology* 19 (1999): 421–430; S.J. Culp et al., "Mutagenicity and Carcinogenicity in Relation to DNA Adduct Formation in Rats Fed Leucomalachite Green," *Mutation Research* 506–507 (2002): 55–63.

320. National Toxicology Program, "Chemicals Nominated to the NTP for In-Depth Toxicological Evaluation or Carcinogenesis Testing in Fiscal Years 1988–1998: Malachite Green" (US Department of Health and Human Services, 9th Report on Carcinogens, 2001). The 10th Report on Carcinogens can be viewed at http://ehp.niehs.nih.gov/roc/toc10.html.

321. M. Milstein, "Imported Seafood Goes Untested: Despite Evidence of Illegal Contaminants in Imported Fish, Only a Tiny Fraction Is Screened before Reaching U.S. Consumers," *The Oregonian,* September 14, 2003, http://www.oregonlive.com/news/oregonian/index.ssf?/base/front_page/1063454304187750.xml.

322. W.M. Grant, *Toxicology of the eye* (Springfield: Charles C Thomas, 1974); T. Bielicky, and M. Novak, "Contact-Group Sensitization to Triphenylymethane Dyes," *Arch. Derm.* 100 (1969): 540–543.

323. National Toxicology Program, "Toxicology and Carcinogenesis Studies of Malachite Green Chloride and Leucomalachite Green in Rats and Mice," National Toxicology Program Technical Report 527 (US Department of Health and Human Services, 2004), http://ntp-server.niehs.nih.gov/meetings/2004/2004FebTRRSMtgpg.html.

324. Ibid.

325. "Malachite Green—Carcinogenic?" *Salmon Farm Monitor,* February 2004,

http://www.salmonfarmmonitor.org/intlnewsfebruary2004.shtml#item9.
326. J. Reynolds, "Contaminated Salmon on Sale to Public" *The Scotsman*, August 5, 2003, http://news.scotsman.com/topics.cfm?id=844672003&tid=15.
327. R. Roberts, "The Lingering Legacy of Malachite Green" *Aquaculture Magazine*, March/April, 2003 http://www.aquaculturemag.com/siteenglish/printed/archives/issues03/203a.html.

Conclusion

328. A. Ross, "Nuvan Use in Salmon Farming: The Antithesis of the Precautionary Principle," *Marine Pollution Bulletin* 20 (1989): 372–374; A. Ross, "UK Usage of Pesticides—Controls and Lessons To Be Learned," in *Interactions between Aquaculture and the Environment*, ed. P. Oliver and E. Colleran (Dublin: An Taisce, 1990); A. Ross, "Leaping in the Dark" (Perth: Scottish Wildlife and Countryside Link, 1997); D. Staniford, "The One That Got Away: Marine Salmon Farming in Scotland" (report for Friends of the Earth Scotland, June 2001), http://www.foe-scotland.org.uk/nation/fish.html.
329. C.J. Redshaw, "Ecotoxicological Risk Assessment of Chemicals Used in Aquaculture: A Regulatory Viewpoint," *Aquaculture Research* 26 (1995): 629–637; "Monitoring of Sea Lice Treatment Chemicals in Southwestern New Brunswick" (Department of Fisheries and Oceans (Canada) Science High Priority Project Final Report, 1995/96); G.K. Rodger, "Development and Application of Biomarker/Bioassay Procedures for the Environmental Monitoring of Sea Lice Treatment Chemicals Used in Salmon Farming—Phase 1" (Aberdeen: Scottish Executive, 1999), http://www.sepa.org.uk/pdf/aquaculture/projects/poster_sea_lice_chemicals_99.pdf; I.M. Davies et al., "Targeted Environmental Monitoring for the Effects of Medicines Used To Treat Sea-Lice Infestation on Farmed Fish," *ICES Journal of Marine Science* 58, no. 2 (2001): 477–485.
330. Natural History Museum, "Environmental Impact of Sea-Lice Treatments," Department of the Environment contract EPG1/5/64 (confidential report for the Department of the Environment, the Veterinary Medicines Directorate, and the Veterinary Products Committee, 1997); "The Post-Authorisation Assessment of the Environmental Impact of Sea-Lice Treatments Used in Farmed Salmon," CTA 9811 (Ministry of Agriculture, Fisheries and Food, 1998); "Ecological Effects of Sea Lice Treatment Agents, (private and confidential report from Scottish Association for Marine Science, April 2001), http://list.zetnet.co.uk/pipermail/seatrout-rev/2002-April/000164.html; R. Edwards, "Big Catch: Fish Farming Is Flourishing at the Expense of Other Marine Life," *New Scientist*, April 17, 2002, http://list.zetnet.co.uk/pipermail/seatrout-rev/2002-April/000164.html; R. Edwards, "Fish Farmers 'Blocked' Vital Safety Study: Salmon Producers and Scientists Furious as Leaked Secret Report Reveals Catalogue of Problems in £4m Pesticide Probe," *Sunday Herald*, April 28, 2002,

http://www.sundayherald.com/24181; "Chemicals Used on Salmon May Hurt Food Chain—Report: Toxic Chemicals Used by Salmon Farms Could Be Killing off Tiny Creatures That Are Vital to the Marine Food Chain, New Scientist Magazine Said Yesterday," Reuters, April 25, 2002. "Leaked Government Report Reveals Fish Farm Damage—Concerns Government Study Being 'Hampered' by Farmers" (release from Friends of the Earth Scotland, April 25, 2002), http://www.foe-scotland.org.uk/press/pr20020405.html.

331. Scottish Association for Marine Science and Napier University, "Review and Synthesis of the Environmental Impacts of Aquaculture" (Edinburgh: Scottish Executive Central Research Unit, 2002), http://www.scotland.gov.uk/cru/kd01/green/reia-00.asp.
332. Wood Mackenzie Veterinary Database (2000) at http://www.woodmac.com.
333. A.N. Grant, "Medicines for Sea Lice," *Pest Management Science* 58 (2002): 521–527.
334. B.M. MacKinnon, "Sea Lice: A Review," *World Aquaculture,* September 1997; K. Boxapen and J.C. Holm, "The Development of Pyrethrum-Based Treatments against Ectoparasitic Salmon Lice in Sea Cage Rearing of Atlantic Salmon," *Aquaculture Research* 32 (2001): 701–707.
335. G.H. Rae, "Sea Louse Control in Scotland, Past and Present," *Pest Management Science* 58 (2002): 515–520.
336. M.W. Jones et al., "Reduced Sensitivity of the Salmon Louse to the Organophosphate Dichlorvos," *Journal of Fish Diseases* 15 (1992): 191–202; G.J. Devine et al., "Chemotherapeutant Resistance in Sea Lice: What Is It and What Can Be Done About It?" *Caligus* 6 (2000): 12–14; I. Denholm et al., "Analysis and Management of Resistance to Chemotherapeutants in Salmon Lice," *Pest Management Science* 58 (2002): 528–536; "A Sea Lice 'Silver Bullet'?" *Salmon Farm Monitor,* May 2003, http://www.salmonfarmmonitor.org/intlnewsmay2003.shtml#item7; "Sea Lice Resistance Management," 2004 article on the Schering Plough website at http://www.myfishpharm.com/aquahome.html.
337. The SEARCH website is at http://www.iacr.bbsrc.ac.uk/pie/search-EU.
338. "Use of Chemicals To Kill Lice Increases—Lice Are Becoming More Resistant to Chemicals, Says Vet Aud Skrudland," Fisheries Information Service, March 17, 2001, http://www.fis.com/fis/worldnews/worldnews.asp?l=e&id=17624.
339. J.D. Smith, "Towards Integrated Management for Sea Lice," *Bulletin of the Aquaculture Association of Canada* 3 (1996): 59–61; P. Andersen and P.G. Kvenseth, "Integrated Lice Management in Mid-Norway," *Caligus* 6 (2000): 6–7.
340. R.A. Schnick, "Trends in International Cooperation for Aquaculture Drug Registration," in *Chemotherapy in Aquaculture: From Theory to Reality,* ed. C. Michel and D.J. Alderman (Paris: Office International des Epizooties, 1992), http://ag.ansc.purdue.edu/aquanic/jsa/aquadrugs/Rozume.htm; R.A. Schnick et al., "Worldwide Aquaculture Drug and Vaccine Registration Progress," *Bulletin of the European Association of Fish Pathologists* 17, no. 6 (1997): 251–260,

http://ag.ansc.purdue.edu/ aquanic/jsa/aquadrugs/publications/ world_drug_progress_9-20-99.htm; R.A. Schnick, "Approval of Drugs and Chemicals for Use by the Aquaculture Industry," *Veterinary and Human Toxicology* 40 (Supplement) (1998): 9–17; R.A. Schnick and P. Smith, "International Harmonisation of Antibacterial Agent Approvals and Susceptibility Testing," *EAFP Bulletin* 19, no. 6 (1999): 293–294, http://ag.ansc.purdue.edu/aquanic/jsa/aquadrugs/Rozume.htm.

341. A.N. Grant, "Medicines for Sea Lice," *Pest Management Science* 58:521-527, 2002.
342. R.A. Schnick et al., "Worldwide Aquaculture Drug and Vaccine Registration Progress," Bulletin of the European Association of Fish Pathologists, 17(6):251-260, http://ag.ansc.purdue.edu/aquanic/jsa/ aquadrugs/ publications/world_drug_progress_9-20-99.htm.
343 "Collection and Treatment of Waste Chemotherapeutants and the Use of Enclosed-Cage Systems in Salmon Aquaculture," SNIFFER/SEPA (Stirling: Scottish Environment Protection Agency, 1998), http://www.fwr.org/fisherie/sr9705f.htm.
344. Report by G2 Consulting 2000, cited in D. Staniford, "A Big Fish in a Small Pond: The Global Environmental and Public Health Threat of Sea Cage Fish Farming" (paper presented at a conference in Chile organized by Terram, June 5–6, 2002), http://www.watershed-watch.org/ww/publications/sf/ BigFishSmallPond(Chile).pdf.
345. F. Cameron, "Planned Product Could Keep Medicines out of the Environment—A Scottish Technology Company Has Won a Smart Scotland Award for Its Idea of a Ground-Breaking Product To Keep Unused Fish Health Products out of the Environment," Intrafish, February 3, 2004, http://www.intrafish.com/ article.php?articleID =41767&s=.
346. "The Story of Silent Spring: How a Courageous Woman Took on the Chemical Industry and Raised Important Questions about Humankind's Impact on Nature (release from Natural Resource Defense Council, April 16, 1997), http://www.nrdc.org/health/pesticides/hcarson.asp.

Dying of Salmon Farming

1. M. Sakai, Y. Goma, N. Seto, S. Atsuta and M. Kobayashi, "Experimental Mortalities of Chum Salmon *(Oncorhynchus keta)* Caused by Bacterial Kidney Disease in Freshwater and Sea Water," *Bulletin of European Association of Fish Pathology* 12 (1992): 87–89.
2. B.O. Johnsen and A.J. Jensen, "The Spread of Furunculosis in Salmonids in Norwegian Rivers," *Journal of Fish Biology* 45 (1994): 47–55.
3. T. Needham, "Management of Furunculosis in Sea Cages," *Bulletin of Aquaculture Association Canada* 95 (1995): 3.
4. "Erythromycin Warning," *Streamtips* (DFO publication), summer 1992.
5. "Canadian Response to USA Concerns with Policy on Importation of Atlantic Salmon to British Columbia," memo, March 16, 1992.
6. A.B. Morton and H.K. Symonds, "Displacement of *Orcinus orca* by high amplitude sound in British Columbia, Canada," ICES *Journal of Marine Science* (2002).
7. T.A. Bakke, and P.D. Harris, "Diseases and Parasites in Wild Atlantic Salmon *(Salmo salar)* Populations," *Canadian Journal of Fisheries and Aquatic Sciences* 55 (Suppl.1) (1998): 247–266.

Index